Atmospheric Processes and Systems

The study of atmospheric processes and systems is fundamental to any basic understanding of the nature and functioning of the total physical environment. Such knowledge also provides further insight into extraordinary atmospheric events such as *El Niño*, which is thought to be responsible for the increasing occurrence of natural disasters in the 'Pacific Rim'.

Atmospheric Processes and Systems presents a concise introduction to the atmosphere and the fundamentals of weather. Examining different aspects of the mass, energy and circulation systems observed in the atmosphere, this book provides detailed yet clearly explained accounts of specific phenomena, including the composition and structure of the atmosphere; energy transfers that take place in the atmosphere and at the Earth's surface; the cycle of atmospheric water in terms of evaporation, condensation and precipitation; pressure and winds at the primary or global scale, including upper atmosphere circulations and upper air/surface linkages; secondary air masses and fronts, weather disturbances and local winds. Each of the sixteen chapters includes a detailed case study.

Illustrated throughout with informative diagrams and photos, **Atmospheric Processes and Systems** presents a non-technical introduction to complex themes and processes of the atmosphere, which play such a dominant role in shaping our physical environment and in controlling activities and responses in the cultural environment.

Russell Thompson is Senior Lecturer in Climatology at the University of Reading. Routledge Introductions to Environment Series, edited by Rita Gardner and Antionette Mannion.

Geography/Environmental Science

Routledge Introductions to Environment Series

Published and Forthcoming Titles

Titles under Series Editors:
Rita Gardner and Antoinette Mannion

Environmental Science texts

Environmental Biology
Environmental Chemistry and Physics
Environmental Geology
Environmental Engineering
Environmental Archaeology
Atmospheric Processes and Systems
Hydrological Systems
Oceanic Systems
Coastal Systems
Fluvial Systems
Soil Systems
Glacial Systems
Ecosystems
Landscape Systems

Titles under Series Editors:
David Pepper and Phil O'Keefe

Environment and Society texts

Environment and Economics
Environment and Politics
Environment and Law
Environment and Philosophy
Environment and Planning
Environment and Social Theory
Environment and Political Theory
Business and Environment

Key Environmental Topics texts

Biodiversity and Conservation
Environmental Hazards
Natural Environmental Change
Environmental Monitoring
Climatic Change
Land Use and Abuse
Water Resources
Pollution
Waste and the Environment
Energy Resources
Agriculture
Wetland Environments

Energy, Society and Environment
Environmental Sustainability
Gender and Environment
Environment and Society
Tourism and Environment
Environmental Management
Environmental Values
Representations of the Environment
Environment and Health
Environmental Movements
History of Environmental Ideas
Environment and Technology
Environment and the City
Case Studies for Environmental Studies

Routledge Introductions to Environment

Atmospheric Processes and Systems

Russell D. Thompson

London and New York

First published 1998
by Routledge
11 New Fetter Lane, London EC4P 4EE

Simultaneously published in the USA and Canada
by Routledge
29 West 35th Street, New York, NY 10001

Typeset in Times and Franklin Gothic by Keystroke, Jacaranda Lodge, Wolverhampton
Printed and bound in Great Britain by T.J. International Ltd, Padstow, Cornwall

British Library Cataloguing in Publication Data
A catalogue record for this book is available from the British Library

Library of Congress Cataloging in Publication Data
Thompson, Russell D.
  Atmospheric processes and systems / Russell D. Thompson.
    p.   cm. – (Routledge introductions to environment series)
  Includes bibliographical references and index.
  1. Atmospheric physics. 2. Dynamic meteorology. I. Title.
  II. Series.
  QC861.2.T46   1998
  551.5–dc21   97–51936

ISBN 0–415–17145–8 (hbk)
ISBN 0–415–17146–6 (pbk)

To Gaynor, Sarah and Vanessa, who have endured my atmospheric interests over our years together

Contents

Series Editors' Preface
Environmental Science titles

The last few years have witnessed tremendous changes in the syllabi of environmentally related courses at Advanced Level and in tertiary education. Moreover, there have been major alterations in the way degree and diploma courses are organised in colleges and universities. Syllabus changes reflect the increasing interest in environmental issues, their significance in a political context and their increasing relevance in everyday life. Consequently, the 'environment' has become a focus not only in courses traditionally concerned with geography, environmental science and ecology but also in agriculture, economics, politics, law, sociology, chemistry, physics, biology and philosophy. Simultaneously, changes in course organisation have occurred in order to facilitate both generalisation and specialisation; increasing flexibility within and between institutions in encouraging diversification and especially the facilitation of teaching via modularisation. The latter involves the compartmentalisation of information, which is presented in short, concentrated courses that, on the one hand are self-contained but which, on the other hand, are related to prerequisite parallel and/or advanced modules.

These innovations in curricula and their organisation have caused teachers, academics and publishers to reappraise the style and content of published works. While many traditionally styled texts dealing with a well-defined discipline, e.g. physical geography or ecology, remain apposite there is a mounting demand for short, concise and specifically focused texts suitable for modular degree/diploma courses. In order to accommodate these needs Routledge has devised the Environment Series, which comprises Environmental Science and Environmental Studies. The former broadly encompasses subject matter which pertains to the nature and operation of the environment and the latter concerns the human dimension as a dominant force within, and a recipient of, environmental processes and change. Although this distinction is made, it is purely arbitrary and for practical rather than theoretical purposes; it does not deny the holistic nature of the environment and its all-pervading significance. Indeed, every effort has been made by authors to refer to such interrelationships and provide information to expedite further study.

This series is intended to fire the enthusiasm of students and their teachers/lecturers. Each text is well illustrated and numerous case studies are provided to underpin general theory. Further reading is also furnished to assist those who wish to reinforce and extend their studies. The authors, editors and publishers have made every effort to provide a series of exciting and innovative texts that will not only offer invaluable learning resources and supply a teaching manual but also act as a source of inspiration.

A. M. Mannion and Rita Gardner

1997

Series International Advisory Board

Australasia: Dr Curson and Dr Mitchell, Macquarie University

North America: Professor L. Lewis, Clark University; Professor L. Rubinoff, Trent University

Europe: Professor P. Glasbergen, University of Utrecht; Professor van Dam-Mieras, Open University, The Netherlands

Plates

Figures

Tables

Boxes

Case studies

 # Preface

An introductory textbook on atmospheric processes and systems represents a vital contribution to the Routledge *Introductions to Environment* series, since the elements and weather regimes involved are fundamental to the nature and functioning of the total physical environment. Despite these rather deterministic overtones, it is apparent that the atmosphere plays a dominant role in shaping the physical environment and in controlling activities and responses in the cultural environment.

For example, even as this Preface is being written (October 1997), the Pacific 'rim' is being ravaged by the build-up of a very strong climatic event known as El Niño. This has caused a severe drought in normally monsoonal Indonesia, which has allowed forest fires (deliberately started by commercial plantation owners) to burn out of control for months on end. Consequently, smoke pollution has become an ecological and social disaster in a vast area extending from Sumatra to southern Thailand and the Philippines. At the same time, above-average sea surface temperatures (i.e., El Niño) in the eastern Pacific have 'spawned' Hurricane Pauline, which caused unprecedented death and destruction along the coast of southern Mexico (and especially at Acapulco).

Since this volume is only a very small part of a series covering a wide range of environmental functions and human dimensions, it is important to provide background atmospheric material that is written in a non-mathematical, more understandable way. Consequently, the explanations are graphical in nature, with the underpinning mathematics and physics covered in a non-technical way. Similarly, the illustrations are designed to be helpful and non-technical, with a large number of idealised or schematic figures included to, hopefully, clarify and explain very complex themes and situations. The study of the atmospheric environment by non-meteorologists has been hampered by the scientific content involved and the degrees of integration of the many elements, processes and systems required. It is hoped that the specifically focused material presented here will be understood by a more general (and even non-scientific) readership and (equally as important) applied to the other volumes included in this valuable series.

The author has lectured for 32 years on atmospheric processes and systems to geography undergraduates at universities in England, Australia, New Zealand, Fiji and

Canada, and has appreciated the difficulties faced in teaching primary circulation processes and systems in particular. These problems are related to the fact that wind and pressure patterns in the lower and upper troposphere are clearly interrelated in a complex three-dimensional way. Consequently, the 'division' of material used in Part IV is based largely on convenience, experience and individual preference (which also applies to the order of presentation in the other parts as well). It must be noted here that the word 'Systems' in the title is used purely in the generic sense to indicate important atmospheric linkages and interrelationships. It does not relate to the spirit and purpose of general systems theory, *viz.* a structured set of objects and attributes, which would only lead to more jargon and confusion.

Dr Russell Thompson
Reading
October 1997

Acknowledgements

The production of this book was greatly facilitated by the assistance and cooperation of the following people:

Professor Michael Breheny, Head of the Geography Department at the University of Reading, for the use of facilities and materials; Heather Browning, Department of Geography, the University of Reading, for her support and cartographic expertise; Kim Butler, Julia Shepherd and Imogen Clarke, who provided the skills and patience of invaluable typists.

The following have kindly given permission for the use of copyright material:

Figures 2.11 and 3.5 with permission of David Higham Associates Ltd; Figures 3.1, 3.3, 9.7(a) and 15.2(c) with permission of Addison Wesley Longman; Figures 3.2 and 4.2 with permission of the University of Chicago Press; Figures 3.4, 9.7, 10.10, 11.15, 15.1(b) and 15.2(a) and (b) with permission of the authors and Routledge; Figures 16.1(b) and 16.2 with permission of the author and Methuen &Company; Figure 10.12 with permission of Chapman & Hall (George Allen & Unwin); Figures 11.13, 16.4(a), 16.5(c) and 16.6(b) with permission of the Royal Meteorological Society; Figure 16.4(b) with permission of the McGraw-Hill Companies.

Part I   Introduction to the Atmosphere

Since this book is concerned with physical processes and interactive systems within the global atmosphere, it is logical to consider first the composition and structure of the atmospheric environment. The word atmosphere is derived from the classical Greek words *atmos* (meaning vapour) and *sphaira* (meaning sphere). However, it is now used less restrictively to denote the gaseous sphere that surrounds Planet Earth and includes water vapour, numerous gases and aerosols.

1 The composition of the atmosphere

Atmospheric constituents have a significant role to play in the functioning of weather systems and short-term climate change. This chapter covers:

- **the average composition of a dry atmosphere**
- **the contribution of variable gases to total atmospheric composition**
- **the contributions of carbon dioxide and ozone**
- **the roles of water vapour and particulate matter**
- **case study: the acid rain problem**

Air (the material of which the atmosphere is composed) is a mechanical mixture of a number of different gases, each of which acts independently of the others. In determining the relative proportions of these 'constant' constituents, it is convenient to consider them first in terms of dry air, which is free from any variable components both in space and time (particularly water vapour and carbon dioxide). A typical analysis gives the percentages listed in Table 1.1. These three gases make up 99.96 per cent of dry air, and the remaining four-hundredths of one per cent consist of minute quantities of various inert gases (namely neon, helium, krypton and xenon), hydrogen (H_2), ozone (O_3), carbon dioxide (CO_2), methane (CH_4), halogen derivatives of organochlorine compounds such as chlorofluorocarbons (CFCs), nitrous oxide (N_2O) and aerosols. With the exception of O_3, CO_2, CH_4, CFCs and aerosols, the components are all 'fixed' gases that do not vary in their relative amounts up to a height of 80 km. Consequently, near the surface, there is a balance between the production and destruction of these gases, especially nitrogen and oxygen.

For example, nitrogen is removed from the atmosphere primarily by biological processes (mainly involving soil bacteria) and is replaced by decaying plant and animal matter. Oxygen is removed from the atmosphere when organic matter decomposes and when oxidation takes place with a wide range of substances. Its removal is also associated with respiration, when the lungs take in oxygen and release carbon dioxide. Conversely, oxygen is added to the atmosphere during photosynthesis by

Table 1.1 *Average composition of the dry atmosphere (below 80 km)*

Component	Symbol	Volume %	Weight %
Nitrogen	N_2	78.08	75.51
Oxygen	O_2	20.95	23.15
Argon	Ar	0.93	1.28

Table 1.2 *Contribution of variable gases to the total atmospheric composition*

Gas (and particles)	Symbol	Volume %	Concentration/parts per million (ppm)
Water vapour	H_2O	0 to 4	
Carbon dioxide	CO_2	0.035	355
Methane	CH_4	0.00017	1.7
Nitrous oxide	N_2O	0.00003	0.3
Ozone	O_3	0.000004	0.04
Particles/aerosols (dust, sulphates, etc.)		0.000001	0.01
Chlorofluorocarbons	CFCs	0.00000001	0.0001

Source: Ahrens, 1991

plants when solar radiation is used to combine carbon dioxide and water to produce sugar and oxygen. In terms of the functioning of atmospheric processes and systems interactions, these fixed gases have little significance. Far more important are the variable and minute ('trace') gases (such as carbon dioxide, nitrous oxide, and methane) and aerosols, CFCs, ozone and (of course) water vapour. Table 1.2 illustrates the percentage composition of these variable gases, and their minuscule contribution to the total concentration masks the fact that these gases are responsible for large-scale global warming/cooling events (see Case Study 3). Furthermore, the important role of CFCs in ozone depletion (see Case Study 2) cannot be appreciated from Table 1.2, where their 0.0001 ppm concentration is the smallest of all the constituents represented.

It is apparent that variable trace gases have a significant role to play in the functioning of weather systems and short-term climate change. The contribution of carbon dioxide to the so-called greenhouse effect and global warming has received considerable media attention in recent years, and the contributions of methane, nitrous oxide and CFCs are also significant (see Case Study 3). The so-called ozone 'hole' over Antarctica (see Case Study 2) has been public knowledge since 1984, and the banning of CFCs resulted from universal concern about increasing skin cancer. Carbon dioxide and ozone are clearly the most important variable gases in the atmospheric environment and their supply/removal from the atmosphere demands special consideration.

Carbon dioxide is a product of combustion, soil processes, oceanic evaporation and various organic processes (e.g. respiration) and is continuously consumed by plants through photosynthesis. Although these processes are not always balanced, the oceans try to regulate the supply of carbon dioxide by dissolving considerable amounts. Even so, carbon dioxide variations are evident over time associated with the 'locking-up' of large quantities of carbon in the form of coal and oil (e.g. 10^{14} tons of CO_2 were withdrawn from the atmosphere during upper Carboniferous coal deposition 300 million years ago). Conversely, this carbon can be liberated (and released into the atmosphere) by the burning of fossil fuels. For example, it has been revealed that the

CO_2 content of the atmosphere has increased by 18 per cent since 1900 (from 296 ppm to 364 ppm in 1997).

It is now generally accepted that these significant changes in carbon dioxide concentrations have a dramatic role to play in the functioning of the greenhouse effect, through the absorption and re-radiation back to Earth of long-wave infrared terrestrial radiation. The exact mechanisms and consequences of this counter-radiation will be discussed in the Chapter 3 Case Study, and it will suffice to say here that CO_2 depletion/accumulation are associated with global cooling/warming episodes. Carbon dioxide is essential for life on Planet Earth, for without the greenhouse effect, the Earth's surface would be 30–40°C cooler and would resemble the surface of the lifeless Moon.

Ozone is concentrated in the stratosphere (Figure 1.1), especially at elevations between 16 km and 25 km, and is formed by the reaction of diatomic oxygen (O_2) with ultraviolet (UV) solar radiation, with wavelengths below 190 nanometres (nm). This solar energy breaks the bond between the two atoms in the diatomic oxygen molecule, and some of the free oxygen atoms (O) collide and bind with a 'normal' oxygen molecule to form triatomic oxygen (O_3) called ozone. This formation is speeded up by a neutral molecule (M), usually nitrogen, which acts as a catalyst for the reaction, and can be represented by the following equations:

$$O_2 + UV \rightarrow O + O \tag{1}$$

$$O + O_2 + M \rightarrow O_3 + M \tag{2}$$

Since molecule M also takes up the kinetic energy released in the above reaction, it accelerates and becomes hotter, so warming the stratosphere (Figure 1.1). This warming trend with increasing elevation is known as a temperature inversion, which is discussed in detail in the following chapter.

The build-up of ozone is further balanced by its natural destruction, since the gas reacts with ultraviolet solar radiation (at longer wavelengths than those required for its formation, between

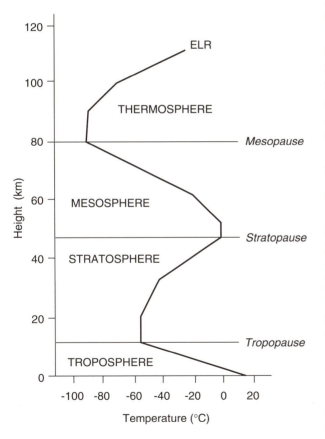

Figure 1.1 *The structure of the atmosphere and the distribution of temperature (ELR).*

230 nm and 290 nm), nitric oxide (NO) and nitrogen dioxide (NO_2), as represented in the following equations:

$$O_3 + UV \rightarrow O + O_2 \tag{3}$$

$$NO + O_3 \rightarrow NO_2 + O_2 \tag{4}$$

$$NO_2 + O \rightarrow NO + O_2 \tag{5}$$

Most of the free oxygen atoms, illustrated in equation (3), combine with oxygen and neutral molecules to recreate more ozone, as in equation (2). This maintains the stratospheric chemistry in a state of equilibrium but it accentuates stratospheric warming with the absorption of more UV radiation in the recreation of ozone. Conversely, the chemical reactions in equations (4) and (5) permanently destroy ozone with its conversion into nitrogen compounds. Furthermore, injections of chlorine into the stratosphere from massive volcanic eruptions (such as El Chichón in Mexico in April 1982) can also deplete stratospheric ozone. Chlorine atoms (C1) play exactly the same scavenging role as nitrogen compounds (in equations (4) and (5)), as is evident from the following equations:

$$Cl + O_3 \rightarrow ClO + O_2 \tag{6}$$

$$ClO + O \rightarrow Cl + O_2 \tag{7}$$

The largest natural source of atmospheric chlorine is the production of methyl chloride from burning plant material in forest and grassland fires following lightning strikes. Even though only 10 per cent of this chlorine reaches the stratosphere, calculations made in the 1970s revealed that, at that time, methyl chloride may have been as effective at destroying ozone as were anthropogenically produced chlorofluorocarbons (CFCs). However, it is apparent that current ozone depletion over Antarctica, associated with abnormally high chlorine concentrations, is far too great to be caused by natural sources alone. It is now generally accepted that ozone equilibrium levels are being changed inadvertently by human activities, with a serious depletion in ozone concentrations and the creation of the infamous Antarctic ozone 'hole'. CFCs, which degrade into chlorine, are held responsible for this creation, which will be discussed in the Chapter 2 Case Study.

So far in this chapter, we have emphasised the chemical composition of dry air, especially the contributions of the important variable gases CO_2 and O_3. However, it must be remembered that the atmosphere also contains a significant, if variable, amount of water vapour, with the variability mainly due to global temperature and associated moisture capacity differences between the poles and the equator. Part III will clearly demonstrate that the capacity of the air to 'hold' given amounts of water vapour depends on its temperature. In simple terms, the higher the temperature, the greater the amount of water vapour retained, although Chapter 7 will remind us of the more pertinent saturation vapour content complexities. Water vapour also varies spatially across the Earth's surface associated with the distribution of land and sea masses and marked differences in the rates of evaporation/evapotranspiration (see Chapter 6) and precipitation (see Chapter 8).

A general, global average for the amount of water vapour in the atmosphere is 1 per cent moist air volume, but hot, humid equatorial air can contain as much as 4 per cent water vapour. Despite the fact that the molecular weight of water vapour is less than that of other gases in the air, 90 per cent of the vapour is concentrated within a few kilometres of the Earth's surface. This is due to the increasing remoteness from the sources of atmospheric water (namely the world's oceans, lakes and vegetation) with increasing elevation and the fact that air temperatures aloft are too low to maintain water in its gaseous state. Water vapour is vital in the atmosphere, since it initiates the hydrological cycle (see Chapter 6 and Figure 6.1), maintains the water balance and has a significant role to play in the greenhouse effect (see Case Study 3).

In addition to the various gases mentioned above, the atmosphere contains a considerable amount of particulate matter or aerosols that are held in suspension, especially salt, dust and sulphates, from a wide range of primary and secondary, natural and anthropogenic sources (Table 1.3). Table 1.3 confirms the dominance of the natural supply sources, especially salt, dust and sulphates from hydrogen

Table 1.3 *Aerosol production estimates (10^9 Kg yr^{-1})*

Sources	All sizes	<5μm radius
Natural		
Primary production		
Sea salt	300–100	500
Windblown dust	70–500	250
Volcanic emissions	4–150	25
Forest fires	3–150	5
Secondary production		
(i.e. gas → particle)		
Sulphates from H_2S	37–420	335
Nitrates from NO_2	75–700	60
Converted plant hydrocarbons	75–1095	75
Total natural:	501–4025	1250
Anthropogenic		
Primary production		
Transportation	2.2	1.8
Stationary combustion	43.3	9.6
Industrial processes	56.4	12.4
Solid waste disposal	2.4	0.4
Miscellaneous	28.8	5.4
Secondary production		
(i.e. gas → particle)		
Sulphate from SO_2	110–220	200
Nitrates from NO_x	23–40	35
Converted	15–90	15
Total anthropogenic:	185–483	280
Overall total:	686–4508	1530

Source: Bridgman, 1990; Barry and Chorley, 1998.

sulphide (H_2S), which contribute up to 89 per cent of the total atmospheric aerosol production. Indeed, the only significant anthropogenic contribution is associated with the secondary production of sulphates from sulphur dioxide (produced from the combustion of coal and oil), which is at the centre of the acid rain problem (see Case Study 1 at the end of this chapter).

Box 1.1

The role of impurities in the functioning of weather processes

1   Aerosols are too microscopic to be observed by the naked eye but they collectively influence visibility as haze.

2   They also colour distant objects and through the evening aerosol haze, dark distant objects appear blue and light clouds appear yellow.

3   Aerosols also influence the turbidity or reduced transparency of the atmosphere through the so-called dust-veil effect (see Chapter 3).

4   Aerosols scatter solar radiation back to space and to the Earth's surface as diffuse radiation or skylight (see q in Figure 3.4 and Chapter 3), and this loss to space is associated with global cooling episodes at the Earth's surface.

5   Certain particulates (especially salt and sulphur compounds) are vital for the condensation process (see Chapter 7) since they act as the condensation/hygroscopic nuclei required to effect the change from vapour to liquid or frozen hydrometeors.

CASE STUDY 1: The acid rain problem

Acid deposition, more generally known as acid rain, consists of rainwater, fog droplets or dry matter that is contaminated by natural and anthropogenic aerosols (especially sulphates and nitrates, see Table 1.3) and is precipitated on to the Earth's surface. This deposition represents a serious environmental problem in north-eastern North America, northern Europe and (more recently) in south-east China, particularly since the pollutants are transported up to several thousands of kilometres from the source regions by the prevailing air flow. The best example of this transfer is the acid deposition over southern Norway and Sweden due to south-westerly wind transport from British power stations and road traffic. Rainwater acidity generally creates far greater problems than dry deposition acidity, since the pollutants concerned react with water vapour and in water droplets. They are also generally transported over much greater (continental-scale) distances compared with the more in situ deposition of dry matter, (as shown in Figure 1.2).

Acidity in rainwater is measured as pH on a logarithmic scale of 14 (most alkaline) to 1 (most acid). Natural rainwater has a pH of between 4.8 and 5.6, which is a weak form of carbonic acid from the reaction of atmospheric CO_2 and water. However, combustion of coal and oil introduces pollutants into the lower to middle troposphere,

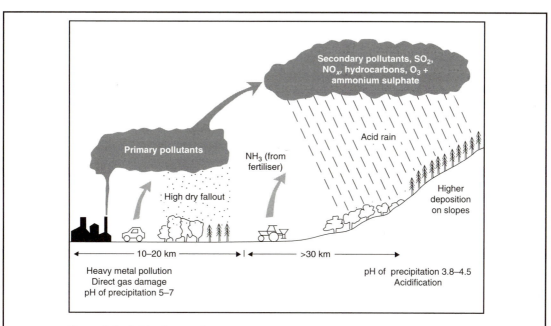

Figure 1.2 *Acid rain transfers.*

such as sulphur dioxide (SO_2) and nitrogen dioxide (NO_2). These emissions now take place at greater altitudes following the UK Clean Air Act of 1956, when tall industrial and power plant stacks were introduced to avoid *in situ* fumigation-type pollution below the inversion 'lid' (see Case Study 5). Consequently, the emission points now coincide with strong middle tropospheric airflow, which facilitates the long-range transport of pollutants over thousands of kilometres. Furthermore, a residence time of up to several days allows the pollutants to react with the water droplets to create sulphuric acid (H_2SO_4) and nitric acid (HNO_3), with pH values below 3 (to resemble vinegar, as at Pitlochrie in Scotland, where a pH of 2.4 was recorded). Table 1.4 illustrates examples of average rainwater (pH) acidity for background (i.e. non-polluted) and polluted locations around the world. Also included are comparative concentrations of sulphate (SO_4^{2-}) and nitrate (NO_3^-) (the major acid ions that control chemical reactions) and ammonium (NH_4^+) which acts as a buffer (to neutralise the acidity).

Reference to Table 1.4 confirms that these ion concentrations are low and small-ranging at the background sites but are considerably higher and wider-ranging at polluted sites, even though four of the five pH values are close together. The pH for Beijing is unusual (only very weakly acidic), given the high ion concentrations, due to the high buffering from NH_4^+ and calcium (460 μ eql^{-1}), which neutralises the acidity.

The environmental consequences of acid deposition are well-documented, especially in southern Norway and Sweden. Here, increased acidity in freshwater lakes (i.e. 2.3 g m^{-2} yr^{-1} deposited, compared with a 'safe' load of 0.3 g m^{-2} yr^{-1}) and streams has decimated fish populations (in 13,000 km^2 of Norwegian lakes) and most forms of biological life. However, in both these examples, the contributions of acid granitic bedrock and acid coniferous litter must not be ignored. Acid deposition also depletes the soil of base cations, which leads to nutrient deficiency (especially magnesium and potassium) and increases the release of toxic ions, such as aluminium. Forests can be seriously depleted, since acid deposition reduces most growth, and ecosystems generally are affected by increased nitrogen uptake and increased acidification of runoff and groundwater recharge. Biodiversity is also reduced in severely stressed ecosystems, where acid-tolerant species flourish.

Table 1.4 *Examples of pH, representing acidity, and concentrations of sulphate, nitrate and ammonium ions in rainwater (in μ eq 1^{-1}), at background and polluted sites around the world*

Location	pH	SO_4^{2-}	NO_3^-	NH_4^+
Background				
Cape Grim, Australia	5.99	4.0	5.0	2.0
New Plymouth, New Zealand	5.57	10.0	3.1	1.5
Amsterdam Island	4.92	8.8	1.7	2.1
Poker Flat, Alaska	4.96	7.1	1.9	1.1
Polluted				
Ithaca, NY, USA	4.16	61.4	29.1	15.7
Sudbury Basin, Canada	4.15	70.8	42.9	25.0
Southern Sweden	4.40	71.0	41.0	42.0
Hradek, Czech Republic	4.40	104.2	198.5	73.6
Beijing, China	6.54	162.5	33.9	160.0

Source: Bridgman (1997).

2 The structure of the atmosphere

Four major zones characterise the atmosphere, based on its thermal properties, although they are not separated with equal degrees of sharpness and do not occur at fixed elevations. This chapter covers:

- **the characteristics of each of the four major zones**
- **the temperature changes within each zone, termed the environmental lapse rate (ELR)**
- **case study: ozone depletion by human activities**

Box 2.1

The weight and pressure of the atmosphere

1   Air has a definite weight, namely 1 m^3 air weighs 113 g.

2   Consequently, the atmosphere exerts a pressure on the Earth's surface amounting to 1.2 kg m^{-3}.

3   In fact, a column of air from the Earth's surface to the top of the atmosphere exerts a pressure that is equivalent to a column of water 11 m high or mercury 760 mm high.

4   This is the reason why mercury barometers (see Chapter 9) are the most realistic and accurate way of measuring variations in atmospheric pressure.

5   Atmospheric pressure decreases with height at a rate of 6 cm 1000 m^{-1}, since it depends on the weight of the air above it.

6   Assuming a normal distribution of temperature, pressure simply becomes a function of altitude and at 5 km, the pressure (0.7 kg m^{-3}) is almost half that at sea level (1.2 kg m^{-3}).

Four major atmospheric zones are generally recognised (see Figure 1.1) based on the thermal properties of the atmosphere, although they are not separated with equal degrees of sharpness and do not occur at fixed elevations. The outer limit of the atmosphere is practically impossible to define, although it may be calculated as about 32,000 km, which is the distance at which the Earth's gravitational pull approximates the centrifugal force of the Earth's rotation. However, even though the atmosphere reaches to these great altitudes, it should be noted that 99 per cent of the atmosphere is within 32 km of the Earth's surface. Furthermore, 90 per cent of the water vapour and virtually all of our weather systems occur in the lowest 16 km, in the zone known as the troposphere. This is the zone in which clouds and precipitation form, winds blow, storms develop and all other weather phenomena occur. The troposphere varies in depth both spatially and temporally, being thickest at the equator (18 km) and thinnest at the poles (8 km) and higher in the summer months. Over the UK, for example, the level of the tropopause (i.e. the upper limit of the troposphere) oscillates between about 6 km and 12 km, being generally lowest during the winter period of low temperatures and low pressure.

The tropopause represents a zone of marked temperature change (see Figure 1.1), and up to this level the tropospheric air tends to cool progressively due to increasing remoteness from the Earth's surface (the main source of air heating), apart from the occasional reversals or warming with elevation, which are known as temperature inversions (see Chapter 5). The normal rate of temperature decrease with increasing elevation in the troposphere is termed the environmental lapse rate (ELR), and it averages 0.65 °C 100 m^{-1}. Chapter 5 explains the functioning and roles of the ELR and emphasises that it is indeed a highly variable rate and it can be steepened or inverted on a regular basis. At the tropopause, the normal temperature decrease stops and isothermal conditions exist (i.e. constant temperatures) well into the lower stratosphere. Indeed, this stable cloudless zone a few kilometres above the tropopause provides ideal flying conditions for supersonic jet airliners like Concorde. Isothermal conditions break down at heights above about 26 km, and the middle/upper stratosphere becomes increasingly warmer (see Figure 1.1) to reach 0 °C at 50 km (at the stratopause). This reversal of the normal ELR represents a significant inversion of temperature and is associated with the zone of maximum ozone concentration. As was explained in Chapter 1, this stratospheric warming is controlled by the absorption of kinetic energy by nitrogen molecules, which is energy released in the formation of ozone by solar radiation (i.e. equations (1) and (2)).

The zone above the stratopause is generally known as the mesosphere, although some meteorologists still refer to it as the upper stratosphere. Temperatures once again decrease in this zone (i.e. a normal ELR) until the mesopause at a height of about 80 km (see Figure 1.1), when conditions again become isothermal. The uppermost zone of the thermosphere (sometimes termed the ionosphere) records an increase of temperature with height, and this temperature inversion is generally related to the absorption of UV radiation by molecular oxygen. At elevations of 100 km above the North Pole, rockets have recorded temperatures of 1480 °C. Also located in the thermosphere are the electrically conducting layers called the Kennelly–Heaviside layers. Gases become ionised or dissociated into individual electrically charged particles called ions. These ionised particles reflect electromagnetic waves and act as a 'mirror' for radio waves.

CASE STUDY 2: Ozone depletion by human activities

Stratospheric ozone is vital for life on Earth since it protects living beings from harmful UV-B radiation (with wavelengths 280–320 nm). Ozone depletion and over-exposure to enhanced UV-B radiation can lead to deadly melanoma skin cancer, increased eye cataracts, damaged photosynthetic activity and destruction of near-surface phytoplankton, which represents the basis of the oceanic food chain. The production and destruction of ozone have already been discussed in Chapter 1, and it should be reiterated that the natural equilibrium of the stratospheric ozone chemistry cycle has been severely interrupted by ozone depletion associated with the decay of chlorofluorocarbons (CFCs). These are contained in a wide range of anthropogenic products, including propellants in aerosol sprays (used for personal and domestic purposes), refrigerants and insulating foam packaging (used by fast-food outlets).

Ozone concentrations are measured by a Dobson spectrophotometer and satellite-based instruments and, over the last 15 years, results from these measurements show a decrease in stratospheric ozone levels of about 1 per cent per year since 1979. Estimates suggest that every 1 per cent decrease in ozone should produce a 2 per cent increase in UV-B radiation (and a 4 per cent increase in skin cancer), although the increasing trend in surface UV-B radiation has yet to be established. Indeed, measurements between 1974 and 1985 at eight centres in the USA revealed an unexpected 11 per cent reduction in UV-B radiation at one centre, smaller reductions at four centres and no change at three centres.

The link between increasing CFC production and ozone depletion was first proposed in 1976 by two Harvard University scientists, which led (without proof) to the removal of 1000 million aerosol sprays from drugstores and supermarkets in the USA (called a new environmental scare theory in the UK). Evidence to support this spray-can link was slow to appear, and indeed measurements of a 20 per cent ozone depletion (first noticed at Halley Bay, Antarctica, in 1982) were considered to be strange and a consequence of old spectro-photometer instrumentation. This apparent anomaly was supported by NASA Nimbus 7 satellite observations but was also rejected since the computers had been programmed to ignore any ozone measurements below 180 Dobson units (i.e. the so-called 'hole') as anomalous and erroneous.

However, in October 1984, a new and more carefully calibrated spectrophotometer confirmed ozone depletion over Antarctica and was used to double-check earlier data to reveal that the decline had been accelerating since the late 1970s (Figure 2.1). NASA re-programmed the Nimbus 7 data and confirmed the existence of a springtime ozone 'hole' over the entire Antarctic continent at elevations between 10 km and 24 km. In addition, measurements of CFC concentrations (in the form of organochlorine), which had been made above Halley Bay since the early 1970s, revealed significant increases during this period of ozone depletion (Figure 2.1), confirming a causal relationship first proposed in 1976, and revealed in equations (6) and (7).

International action on ozone depletion was first undertaken by UNEP in April 1980, which called for governments to reduce the production and use of CFCs. A working group was set up to draft a convention for the protection of the ozone layer, which culminated in the Vienna Convention of March 1985 (i.e. two months before the Antarctic ozone 'hole' was first reported in the journal *Nature*). The convention dealt with international co-operation in ozone research but lacked any

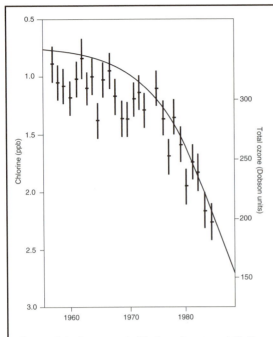

Figure 2.1 *Ozone and chlorine changes at Halley Bay, Antarctica, since 1957 (after Gribbin, 1988).*

specific controls. Indeed, later attempts to formulate a protocol on the reduction of CFCs (in December 1986 and February 1987) failed, partly due to British opposition. However, partly due to the infamous 'U-turn' of the Thatcher Government in Montreal in September 1987, success finally came when 40 countries adopted a CFC reduction protocol (with a 50 per cent reduction target by 1999).

A few weeks after the Montreal Protocol was signed, measurements at Halley Bay revealed that the ozone 'hole' was almost complete at an altitude of 16.5 km on 7 October 1987, following a 97.5 per cent destruction of the amount of ozone recorded on 15 August 1997. This confirmed the seriousness of the situation (not truly appreciated at Montreal) and caused the Montreal document to be updated in 1990 and 1992, accelerating the requirements for phase-out (Table 2.1). Unfortunately, the CFCs already in the atmosphere will have a long residence time (between 75 and 100 years), since they rise only very slowly into the stratosphere (taking 10 years to reach the lower edge of the ozone layer revealed in Figure 1.1). So, despite the revised Copenhagen agreements (Table 2.1), ozone depletion will continue to represent a serious problem in polar regions well into the next millennium. This will be especially so over Antarctica, where the circumpolar vortex is particularly strong (see Chapter 11) and where the stratospheric temperatures and chemistry are most conducive to chlorine scavenging. Substitutes for these substances have been developed and are being tested for potential environmental damage. Some substitutes contain chlorine atoms, such as HCFCs. A slow phase-out of these substitutes is scheduled by 2030, to accommodate the techno-logical developments needed for non-chlorine replacements.

Table 2.1 *Phase-out periods of CFCs, halocarbons and similar substances agreed to in Copenhagen, 1992*

Ozone-depleting substance	Phase-out period
CFC–11, CFC–12, CFC–113, CFC–114, CFC–115	Reduce by 75% by 1994 (from 1986 levels); total phase-out by 1996
Halon-1211, Halon-1301, Halon-2402	Total phase-out by 1994
CCl_4 (carbon tetrachloride)	Reduce by 85% by 1995; total phase-out by 1996
CH_3Br (methyl bromide) (fire extinguishers)	Freeze by 1995; further research needed

Source: Bridgman, 1997.

Key topics for Part I

The composition and structure of the atmosphere provide the framework for wide-ranging processes and interactive systems.

1 Nitrogen and oxygen dominate the atmospheric composition and are represented by distinctive cycles, where the production of these gases is balanced by their destruction.
2 In terms of the functioning of atmospheric processes and systems interactions, these two fixed gases have little significance. Far more important are the variable and minute 'trace' gases, particularly water vapour, greenhouse gases (especially carbon dioxide) and CFCs, which are responsible for large-scale climate change.
3 Water vapour and greenhouse gases are associated with global warming when enhanced and cooling when depleted. Without the natural greenhouse effect, the Earth would be 30–40 °C cooler than at present and indeed would resemble the lifeless lunar surface.
4 CFCs degrade into chlorine compounds, which deplete the stratospheric ozone, especially over the Antarctic continent, where the atmospheric circulation promotes these chemical reactions.
5 Both carbon dioxide and CFCs are released into the atmosphere by human activities and this enhancement must be rigidly controlled.
6 Atmospheric aerosols also represent an atmospheric problem, particularly with long-range transport and acid deposition in lakes and forests far from the source regions.
7 The atmospheric temperature changes (the ELR) take the form of decreases in the troposphere (normally) and mesosphere and increases (inversions) in the stratosphere and thermosphere.

Further reading for Part I

Global Air Pollution. Howard Bridgman. 1990. Belhaven Press.
Provides a detailed, technical introduction to the problems of acid rain and ozone depletion.

The Hole in the Sky. John Gribbin. 1988. Corgi Press.
Provides a general, non-technical account of human beings' threat to the ozone layer.

Global Environmental Issues. David Kemp. 1994. Routledge.
Provides a detailed, non-technical introduction to a wide range of atmospheric problems including acid rain, turbidity and the threat to the ozone layer.

Global Environmental Change. A. M. Mannion. 1997. Addison Wesley Longman.
An up-to-date and authoritative examination of all aspects of environmental change in the Earth–atmosphere system, including acid rain, global warming and ozone depletion.

Part II Radiative Fluxes and Energy Transfers in the Atmosphere and at the Earth's Surface

This section studies the supply of energy to the atmosphere, which initiates and sustains a wide range of atmospheric processes and systems. It will be demonstrated that solar radiation is the ultimate source of energy leading to all of the varied atmospheric conditions experienced over Planet Earth. Solar radiation surpluses and deficiencies create pressure differences within the atmosphere (see Chapter 9), for example equatorial low pressure and polar anticyclones, which cause the winds to blow (see Chapter 10) and foster the formation of precipitation and aridity (see Chapters 8 and 9). The Sun is regarded as the great 'engine' that drives the winds and ocean currents (see Chapter 10), generates the weather (see Part V) and makes Planet Earth a liveable place for human beings.

3 Radiative fluxes and radiation balance

Radiative fluxes and the radiation balance generate all weather processes in the troposphere and vary both spatially and temporally. This chapter covers:

- **the solar/electromagnetic spectrum**
- **the latitudinal variation of solar radiation/insolation**
- **the receipt and disposal of solar radiation within the troposphere**
- **the balancing role of outgoing terrestrial radiation**
- **counter-radiation and the greenhouse effect**
- **case study: the enhanced greenhouse effect due to human activities**

The sole input to energy flows within the atmosphere is radiation emitted from the Sun, 150 million km away from the Earth with a surface temperature of 6000 Kelvin (K). This solar radiation is radiant energy in the form of electromagnetic waves, travelling at the speed of light (3×10^8 m s^{-1}) and spread over a very broad band of wavelengths (Figure 3.1a). This band is generally known as the 'solar spectrum' and is characterised by increasing wavelength (μm) and decreasing frequency (Hz), from gamma rays/X-rays through to so-called 'middle' infrared waves (see Table 3.1 and Figure 3.1), which make up 99.59 per cent of the spectrum.

The amount of solar radiation actually incident at the top of the atmosphere depends on the time of year, time of day and latitude (which controls the obliquity of the solar beam). However, on an annual basis and spread uniformly over the outer edge of the atmosphere, the amount of solar radiation received per unit area and time averages approximately 338 W m^{-2}. This value is only 25 per cent of the solar constant, which represents the maximum possible short-wave radiation receipt at any point in the Earth–atmosphere system. However, the solar constant value can be approached at some low-latitude tropical locations at high elevations when the Sun is directly overhead (at zenith) in cloudless, impurity-free atmospheric conditions (e.g. at the summit of Mount Kenya, at the March/September equinoxes).

At the Earth's surface, the variability in arriving solar radiation (termed insolation) is controlled by latitude. Figure 3.3 clearly illustrates that, due to the greater obliquity of the solar beam with increasing latitude, the annual insolation value in the subtropics is about three times more than that in polar areas. However, Figure 3.3 also reveals anomalously low insolation values in the humid equatorial regions, such as the Congo/Zaire, which are up to 50 per cent lower than in the neighbouring arid

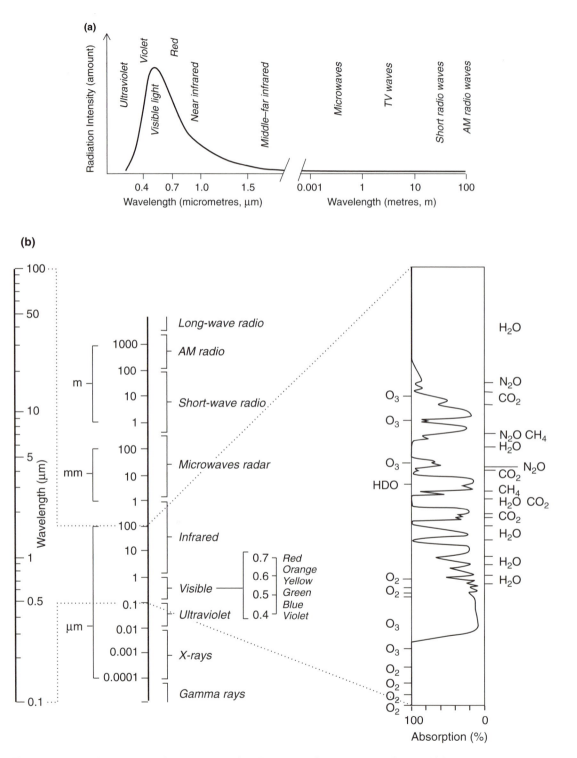

Figure 3.1 *The electromagnetic spectrum (after Henderson-Sellers and Robinson, 1987).*

Table 3.1 *The solar/electromagnetic spectrum (1 micrometre, μm, is one-thousandth of a mm)*

Wavelength (μm)	Approximate frequency (Hz)	Waves		% of spectrum
0.15				
	10^{19}	Gamma rays/X-rays		
0.2				7.29
	10^{15}	Ultraviolet (UV)		
0.4				
	10^{14}	Visible violet	0.42	
		blue	0.48	
		green	0.52	
		yellow	0.56	43.50
		orange	0.60	
		red	0.68	
0.7				
		Near infrared		36.80
1.5				
	10^{13}	Middle infrared		12.00
5.6				
		Far infrared		
1000				
	10^{11}			
>1000		Micro- and radio waves		0.41
	10^{8}			

subtropics, especially around 20° N in the Red Sea area. It is apparent that a significant amount of solar radiation is lost in the atmosphere due to the persistent equatorial cloudiness (see Chapter 9) and high reflection rates (Figure 3.4). It is apparent that during its passage from the top of the atmosphere to the Earth's surface, the solar radiation has to penetrate and diffuse through clouds and a variety of atmospheric constituents such as water vapour, aerosols and numerous gases (see Chapter 1). Consequently, part of the spectrum (Figure 3.1) is absorbed, scattered or reflected and the remainder is transmitted to the Earth's surface as direct solar radiation. Figure 3.4 illustrates this dispersal (and eventual disposal) of solar radiation within the atmosphere, and all individual fluxes are expressed as a percentage of the 338 W m^{-2} solar energy that arrives at the top of the atmosphere (discussed above). It confirms that 25 per cent of the beam is scattered (*S*) by air molecules and minute impurities.

Scattering is a selective process (generally referred to as Rayleigh scattering) in that the shorter blue wavelengths (at about 0.48 μm, see Table 3.1) are scattered more readily than the longer red wavelengths, accounting for the blue colour of the sky.

Box 3.1

The flux and wavelength of solar radiation

1 The rate of flow (flux) of radiation from the Sun is obtained from the Stefan–Boltzman Law, which states that a radiation flux is directly proportional to the fourth power of its absolute temperature.

2 It is expressed in the following equation:

$$\text{Radiation flux} = \sigma T^4 \tag{8}$$

Where σ is the Stefan–Boltzman constant (5.67×10^{-8} W m^{-2} K^{-4}) and T is the Sun's temperature (6000 K).

3 Using this equation, the solar radiation flux is equal to 1370 ±10 W m^{-2} (Figure 3.2), and this is known as the solar constant (i.e. the energy received outside the atmosphere on a plane normal to the solar beam).

4 The wavelength of solar radiation is obtained from the Wien Displacement Law, which states that its wavelength of maximum intensity of emission (λ_{max}) is inversely proportional to the absolute temperature (T).

5 It is expressed in the following equation:

$$\lambda_{max} \, (\mu m) = 2897 \, T^{-1} \tag{9}$$

6 For the sun (at 6000 K) the wavelength of maximum flux is 0.48 μm (Figure 3.2), which occurs in the blue waves of the visible part of the solar spectrum (Table 3.1).

However, with increasing aerosol sizes, as are found over urban/industrial regions, especially with low solar angles, the longer wavelengths can indeed be scattered to provide the spectacular red sunsets over these regions. It should be noted from Figure 3.4 that only a relatively small part (4 per cent) of the scattered radiation is lost to space. It is apparent that 21 per cent is scattered downwards and reaches the Earth's surface as indirect diffuse radiation (q) or skylight (the typical solar radiation receipt on cloudy or overcast days).

Reflection (R) in Figure 3.4 is a non-selective process and affects all wavelengths of the spectrum. It is mainly from larger aerosols and cloud tops (and depends on the cloud's thickness), where the proportion of incident radiation that is reflected (termed the albedo, α) averages 55 per cent and accounts for 23 per cent of the solar beam that is lost into space. As was discussed earlier, high reflection rates from thick cumulus clouds (see Chapter 8) in the humid tropics cause anomalously low insolation rates in these regions. Absorption (A) by clouds (3 per cent) and atmospheric gases (21 per cent) occurs primarily at the shorter wavelengths (below 0.4 μm, see Figure 3.1b) where oxygen (O_2) and ozone (O_3) absorb the bulk of the gamma rays, X-rays and

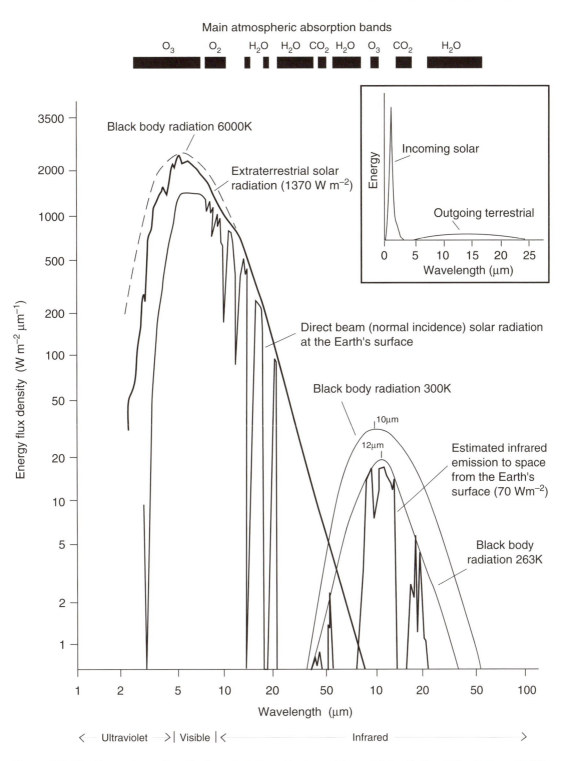

Figure 3.2 *The flux and wavelength characteristics of solar and terrestrial radiation (after Sellers, 1965).*

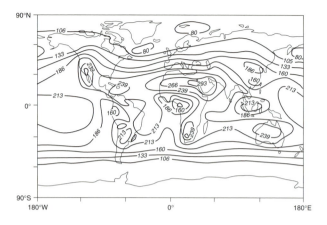

Figure 3.3 *Average annual solar radiation at the Earth's surface (W m⁻²) (after Henderson-Sellers and Robinson, 1987).*

lethal UV radiation (see Chapter 1). Finally, 28 per cent of the solar beam reaches the Earth's surface as direct solar radiation (Q in Figure 3.4). However, 4 units are reflected (R) back to space immediately due to the Earth's albedo, which, as Table 3.2 illustrates, ranges from 0.13 for tropical forests to 0.37 for dry sandy deserts and 0.80 for snow-covered ice. Table 3.2 also reveals that the albedo of a water surface increases as the solar zenith angle increases, and the albedo of an ocean surface also increases with the occurrence of white caps on the waves.

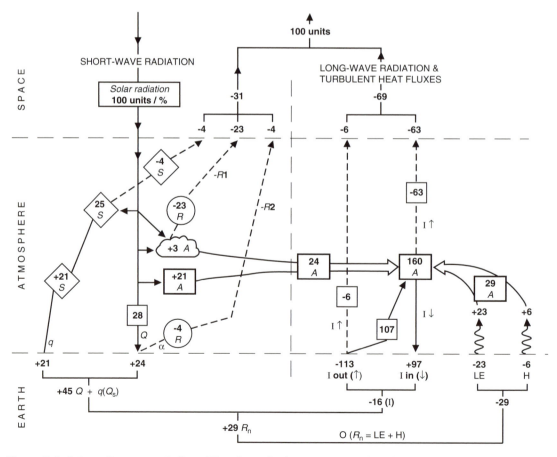

Figure 3.4 *Schematic representation of the atmospheric energy cascade (after Barry and Chorley, 1998).*

Table 3.2 *Albedos and emissivities of common surface types (annual means)*

Type	Albedo α	Emissivity ε
Tropical forest	0.13	0.99
Woodland	0.14	0.98
Farmland/natural grassland	0.20	0.95
Semi-desert/stony desert	0.24	0.92
Dry sandy desert/salt pans	0.37	0.89
Water (0–60°)	<0.08	0.96
Water (60–90°)	<0.10	0.96
Sea ice	0.25–0.60	0.90
Snow-covered vegetation	0.20–0.80	0.88
Snow-covered ice	0.80	0.92

Source: Henderson-Sellers and Robinson, 1987.

From the above discussion (and reference to Figure 3.4), it is apparent that, on average, 31 units of solar radiation are transmitted back to space immediately by scattering (4 units), reflection from clouds/aerosols (23 units) and reflection from the Earth's surface (4 units). Also, 24 units are absorbed temporarily within the atmosphere (by clouds and gases), to be eventually radiated back to space. Therefore, slightly less than half of the solar radiation received at the top of the atmosphere (45 units, or 152 W m^{-2}) actually reaches and is absorbed by the Earth's surface as Q_s or insolation (24 units direct beam, Q and 21 units diffuse skylight, q; Figure 3.4). This energy is converted from radiation into thermal energy, which heats the Earth's surface which, in turn, becomes a source of long-wave radiation (I↑), together with a considerable amount of energy released by the Earth's materials (Figure 3.4). At 293 K, this terrestrial radiation is mostly emitted in the infrared spectral range from 5 to 50 µm (Figure 3.2), with a peak emission at 10.2 µm (using equation 9). Using equation 8, this radiation transfer equals 113 units, which is considerably more than the 45 units of insolation and includes 97 units of continuous counter-radiation, which constitutes the so-called greenhouse effect (see Chapter 1).

Although the atmosphere is mostly transparent to incoming short-wave radiation (absorbing only 24 per cent of the incident solar beam, as in Figure 3.4 and discussed above), it readily absorbs 90 per cent of the emitted long-wave terrestrial radiation. It is worth noting here that the actual emissivity (ε) of the surface varies within a small range, depending on surface temperature (Table 3.2b) from 0.99 for tropical forest to 0.88 for snow-covered vegetation. The principal atmospheric absorbers are water vapour (5.3–7.7 µm and beyond 20 µm), ozone (9.4–9.8 µm), carbon dioxide (13.1–16.9 µm) and clouds (all wavelengths). Consequently, only 6 units of the total terrestrial radiation (113 units) escape directly to space (Figure 3.4), mainly through the atmospheric 'window' (between 8.5 µm and 14.0 µm), where water vapour and carbon dioxide are weak absorbers (Figure 3.2).

However, as was discussed in Chapter 1, anthropogenic CFCs are active absorbers in this 'window' and clearly enhance the greenhouse effect. The rest of the I↑ radiation is absorbed (A) by the atmosphere (107 units), which, in turn, re-radiates the absorbed energy partly into space (10 units) and mainly (97 units) back to the Earth's surface as long-wave returning or counter-radiation. The resultant, vital greenhouse effect considerably reduces the net or effective outgoing long-wave radiation loss from the Earth's surface from what would be observed with a perfectly translucent atmosphere. Furthermore, as discussed earlier, the absence of this counter-radiation would reduce the temperature at the Earth's surface by 30–40 °C, to compare to the lifeless lunar surface (see Chapter 1).

CASE STUDY 3: The enhanced greenhouse effect

The term 'greenhouse effect' is used to describe the way in which certain gases in the atmosphere respond to solar and terrestrial radiation in a way similar to that of the glass panes of a greenhouse. Solar radiation is transmitted through the transparent glass and is then absorbed by Earth-surface objects, converting it into heat and causing a marked temperature rise in the greenhouse. The next stage is when this heat is emitted away from the surface as long-wave infrared radiation, which is absorbed by the (now) opaque glass and re-radiated back to the greenhouse floor. Here, it is absorbed by surface objects, is converted into heat and enhances the heating of the greenhouse.

The atmosphere, however, does not behave in the same way as a greenhouse, since a significant amount of solar radiation is absorbed in the gases and clouds (i.e. the 24 units in Figure 3.4), especially below 0.4 µm as indicated in Figure 3.1b. Furthermore, the trace gases responsible for the absorption of outgoing infrared radiation are ineffective between 8.5 µm and 14.0 µm, although this atmospheric 'window' is analogous to an open window in a glass house. Furthermore, the atmosphere is prone to wind movement (not found in a horticultural greenhouse) and the reduction of heat build-up by convection and advection currents (see Chapter 5). Despite these reservations, the term 'greenhouse effect' is universally applied to the similar processes taking place within the atmosphere and is adopted throughout this book.

Greenhouse gases are (trace) atmospheric constituents that respond particularly to infrared radiation in a similar way to the glass panes of a greenhouse. The origin and importance of these gases were explained in detail in Chapter 1, and this case study will concentrate on their enhancement in recent years and their predicted contributions to global warming in the future. However, it must be noted that despite all the greenhouse euphoria of recent years, this effect is certainly not a new discovery. In the 1860s, atmospheric scientists first realised that trace gases had a role in keeping Planet Earth pleasantly warm. Indeed, in 1863, John Tyndall discussed the role of water vapour in such warming in a paper published in the *Philosophical Magazine*. Also, in 1896, Svante Arrhenius published a paper related to changes likely to be caused by increasing concentrations of atmospheric carbon dioxide (CO_2) from coal combustion.

Since the mid-1970s, the prospect of greenhouse warming of Planet Earth has received widespread attention with the accumulation of evidence from the Mauna Loa observatory in Hawaii, which started measuring CO_2 in 1957 under the direction of the Scripps Institution of Oceanography. At 3400 m, this site is remote from sources of urban/industrial pollution and deforestation and was considered to represent a natural atmosphere free from local anthropogenic contamination. By 1975, data confirmed a steady build-up of CO_2 levels over Mauna Loa (representing the Northern Hemisphere) from 310 ppm (or 0.031 per cent) in 1957 to about 325 ppm eighteen years later (Figure 3.5a and b). This supported increases based on earlier estimates from tree-ring analyses, which suggested a baseline concentration in 1850 (i.e. pre-Industrial Revolution) of 270 ppm. Figure 3.5 clearly illustrates that since 1975, CO_2 levels have increased steadily to reach 359 ppm by 1995 (an increase of 10 per cent). Superimposed on this upward trend is an annual oscillation pattern (Figure 3.5b) due to greater photosynthetic uptake (and CO_2 consumption) in summer than in winter (see Chapter 1). Furthermore, predictive modelling suggests that CO_2 levels will reach 540 ppm by the middle of the next century (i.e. doubling the pre-industrial level).

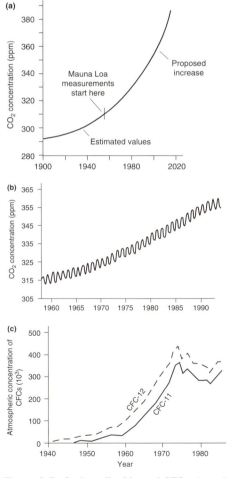

Figure 3.5 *Carbon dioxide and CFCs: twentieth-century changes ((c) after Gribbin, 1988).*

There is no doubt that anthropogenic CO_2 contributions have increased the absorptive strengths for terrestrial infrared radiation in the band between 13 µm and 100 µm. At the same time, inadvertent human activities have increased other important greenhouse trace gases. For example, methane (CH_4) is produced by rice paddies, rubbish tips and livestock populations and has increased from a level of 700×10^9 parts per billion (ppb) in the eighteenth century to about 1670×10^9 ppb today. The atmospheric concentration of methane has increased by 11 per cent over the last decade, and the current annual rate of increase is approaching 1 per cent of the current concentration. Nitrous oxide (N_2O) is produced artificially from chemical fertilisers, oil and coal combustion and as a spray-can propellant and shows a more modest increase of 0.7 ppb per annum (or 0.25 per cent). However, since it can remain in the atmosphere for 150 years or more, its cumulative impact could accentuate global warming for some considerable time. Methane and nitrous oxide contribute 30 per cent of the greenhouse effect caused by trace gases and are effective absorbers of infrared radiation between 7 µm and 13 µm. This range is the so-called atmospheric 'window' where infrared radiation can normally escape into space as both water vapour and CO_2 are ineffective absorbers at these wavelengths.

The final greenhouse constituent worth considering are the CFCs, which first appeared in the 1940s (Figure 3.5c) and are now known to be notorious ozone destroyers (Case Study 2). It is generally accepted that CFCs contribute 15 per cent of the trace-gas greenhouse effect, and they are produced by a wide range of processed compounds used for spray-can propellants, refrigeration, foam-blowing agents and foam hard plastics. CFCs are particularly strong absorbers in the atmospheric 'window' where CO_2 is ineffective, and they are also 20,000 times more efficient in greenhouse activity than CO_2. Figure 3.5c shows that there has been a dramatic rise in the two most commonly produced CFCs, namely CFC-11 (trichlorofluoromethane) and CFC12 (dichlorofluoromethane). However, as was discussed in the Chapter 2 Case Study, production declined in the 1970s following the US ban on spray-cans and especially since the Copenhagen revision of the Montreal Protocol in 1992, when it was agreed to phase out all CFC production by 1996. However, despite these reductions, existing CFC's greenhouse properties will continue to be effective for another century or so due to their remarkable stability over time.

The combined build-up of all these trace gases over the last century or so has contributed to the enhanced greenhouse effect and (probably) the recognised 0.5 °C temperature rise. Future levels of trace-gas concentrations and their impact on global warming have been simulated by (numerical) climate models. These are mostly based on the projected CO_2 doubling (from the 1850 level of 270 ppm) and incorporate equilibrium responses and feedbacks, including water vapour increases, decreases in albedo and cloud changes. Table 3.3 shows the results of four such models published around 1986, which show good agreement between them. The data confirm an annual average global warming of between 3.5 °C and 5.2 °C, with increased global precipitation values between 7 and 15 per cent.

It must be noted that climate models have limitations associated with the simple representation of ocean and cloud characteristics and the omission of stratospheric aerosol loading (see Chapter 1). For example, with Table 3.3, the Met.O model was reused with cloud microphysics properties and variable cloud radiative properties and the global mean warming was reduced from 5.2 °C to 2.7 °C and 1.9 °C, respectively. However,

Table 3.3 *Modelled equilibrium responses for a doubling of atmospheric carbon dioxide levels. Global mean and eastern UK values are given*

Equilibrium response	Model[a]			
	GFDL	GISS	MET.O	NCAR
Warming (°C)				
Global mean	4.0	4.2	5.2	3.5
Eastern UK JJA+	5.0	4.0	5.0	4.0
Eastern UK DJF+	5.5	4.0	6.0	6.0
Precipitation				
Global mean (%)	+9	+11	+15	+7
Eastern UK JJA[b] (mm/day)	−(0–1)	−(0–1)	+(0–1)	+(0–1)
Eastern UK DJF[b] (mm/day)	+(0–1)	+(0–1)	+(0–1)	+(0–1)

Adapted from Rowntree, 1990.
[a] Models are low resolution versions from: Geophysical Fluid Dynamics Laboratory, Princeton, NJ, USA (GFDL); Goddard Institute for Space Studies, New York, USA (GISS); Meteorological Office, Bracknell, UK (MET.O); National Center for Atmospheric Research, Boulder, Colorado, USA (NCAR).
[b] JJA = June, July, August; DJF = December, January, February.

despite the uncertainty observed and the general lack of evidence of a direct cause–effect linkage, some experts argue that recent climate trends are consistent with changes predicted by numerical models to accompany greenhouse warming. These include the occurrence of the ten warmest years on record (i.e. since the 1880s) since 1979; the accelerated recession of mountain glaciers in low latitudes; the massive break-up of ice shelves to the east of the Antarctic Peninsula (e.g. the Larsen Ice Shelf); and the increasing frequency of extreme weather events. Consequently, arguments are put forward at regular intervals (e.g. the Rio Summit in Brazil, 1992, the Berlin Mandate of 1995 and the Rio plus 5 Summit in New York, 1997) so that action can be taken now to counteract enhanced greenhouse warming. These measures include sharp reductions in the consumption of fossil fuels; greater reliance on renewable energy sources (wind power and solar power, for example); energy conservation/improved efficiency; carbon or green taxes; and massive reforestation projects. Figure 3.6 indicates that climate models can produce a wide range of global warming scenarios with changing CO_2 emissions, where scenario C appears to be the most desirable. This, however, can be accomplished only by drastic cuts in emissions over the next decade.

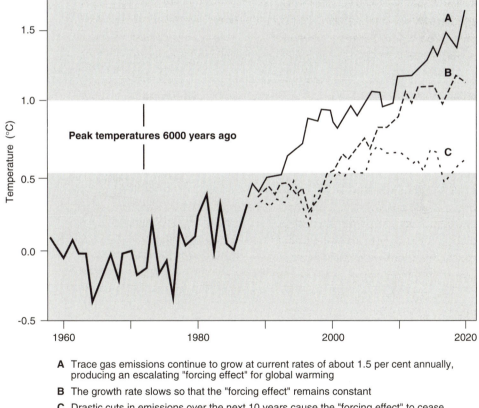

A Trace gas emissions continue to grow at current rates of about 1.5 per cent annually, producing an escalating "forcing effect" for global warming

B The growth rate slows so that the "forcing effect" remains constant

C Drastic cuts in emissions over the next 10 years cause the "forcing effect" to cease growing by 2007

Figure 3.6 *Climate model predictions of global temperature change associated with three carbon dioxide scenarios (after Hansen et al., 1988).*

Net radiation and energy transfers within the atmosphere and at the Earth's surface

The all-wave radiation balance is termed net radiation, which is partitioned into energy (heat) transfers or non-radiative fluxes (termed energy balance). This chapter covers:

- the distribution and disposal of net radiation
- the role of conduction and convection in energy transfers within the atmosphere
- energy transfers at the Earth's surface at the global/continental scale and at particular sites
- case study: radiation balance and energy balance modifications due to urbanisation

Net radiation and its disposal within the troposphere

The previous chapter revealed that the Earth's surface receives 45 units of solar radiation/insolation (Q_s) and loses 16 units of terrestrial radiation (I) (see Figure 3.4), which leaves a 'surplus' of 29 units, which must be disposed of by other (non-radiative) energy transfers. Before this alternative disposal is discussed, it must be noted that the distribution of Q_s and I varies dramatically with latitudinal changes. For example, the equatorial values of Q_s are approximately two and a half times the polar values (i.e. from 237 W m^{-2} at 0° to 98 W m^{-2} at the North Pole). However, the I values are all within 22 W m^{-2}, with the highest losses occurring in the dry cloud-free subtropics (*viz.* 198 W m^{-2}), where the counter-radiation is weakest (*cf* 176 W m^{-2} at the North Pole). Consequently, tropical regions experience a radiation surplus (+48 W m^{-2}), whereas polar areas have distinctive radiation deficits (−78 W m^{-2}). The difference between Q_s and I (or, more correctly all-wave incoming/outgoing radiation) is termed net radiation (R_n), expressed in the following equation (with the terms indicated in Figure 3.4 and explained in the text):

$$R_n = \frac{Q_s}{(Q+q)(I+\alpha)} \pm \frac{I}{I\downarrow - I\uparrow} \tag{10}$$

Chapter 3 (and Figure 3.4) confirms that the Earth has an annual radiant energy surplus or available net radiation of 29 units, which would increase surface temperatures at a rate of 250 °C day^{-1} in the upper centimetre without an effective atmospheric transfer by non-radiative energy fluxes. The required transfer of energy away from the Earth's surface is accomplished by conduction and convection processes. Conduction is the process whereby heat is transmitted within a substance by the collision of rapidly moving molecules. Naturally, this transfer is most effective in solids and least effective in gases. Consequently, molecular conduction is only significant in the atmospheric boundary layer, which represents the 'skin' of air closest to the Earth's surface. It is quite negligible on the atmospheric scale depicted by Figure 3.4.

On the other hand, convection involves the turbulent vertical movement of energy, which can occur only in liquids and gases. In the atmosphere, local air movements or eddies transport energy and water vapour upwards through turbulent motions initiated by free or forced convection. Free convection relates to parcels of air that rise due to a higher temperature and lower density inside the buoyant parcel compared with the surrounding atmospheric environment. Forced or mechanical convection occurs when air parcels are moved aloft over surface obstacles, such as mountain ranges (orographic displacement), or when contrasting air masses are in juxtaposition (frontal uplift). The actual convective mechanisms will be explained in the next chapter, but it is apparent that these processes are responsible for the loss (or partitioning) of the 29 units of net radiation available in Figure 3.4, in the form of both sensible heat (H) and latent heat (LE) fluxes.

However, reference to Figure 3.4 reveals that only 6 out of the 29 units are transferred to (and temporarily stored in) the lower atmosphere (i.e. the boundary layer) by sensible heat fluxes, or direct molecular conduction. Furthermore, this occurs only when the surface is warmer than the lower atmosphere and the sensible heat is released upon mixing with the cooler air. Obviously, the most important convective transfer (involving 23 units) is associated with the evapotranspiration/condensation processes (see Chapters 6 and 7) and the related latent heat fluxes. On average, evapotranspiration from water bodies, soil and plants consumes 23 units of available radiative energy, since 2.45 MJ kg^{-1} is required for the latent heat of vaporisation of water at 20 °C. Furthermore, this energy is stored within the water vapour and is transported to the top of the troposphere (see Figure 1.1) by the free or forced convective systems described earlier. The rising air parcels will eventually cool beyond their dew point levels (see Chapter 7), and the water vapour then condenses to form clouds/precipitation (see Chapters 7 and 8). At this time, the stored energy is released into the troposphere as 23 units of latent heat of condensation (i.e. 2.45 MJ kg^{-1} at 20 °C), which influences the equilibrium or stability tendency of the air (see Chapter 5). It must be noted that these 29 units of sensible heat and latent heat are stored temporarily in the troposphere until they are finally transferred into space by atmospheric radiation (Figure 3.4). These transfers combine with the radiative transfers, explained in Chapter 3 (Figure 3.4), in order to restore the Earth–atmosphere energy balance.

Energy balance at the Earth's surface

So far in this chapter, the partitioning of net radiation has been associated with the roles of sensible and latent heat as contributors to atmospheric convective transfers. It is now time to quantify the actual fluxes concerned and examine the specific environmental controls involved at the Earth's surface. On a global scale, net radiation (Rn) is partitioned between the sensible heat (H) and latent heat (LE) fluxes as follows:

$$R_n = H + LE \qquad (11)$$

However, at this scale, the actual partitioning varies between land and ocean surfaces since, for example, the world's oceans use about 90 per cent of R_n to evaporate water (LE). Conversely, over land, the two heat fluxes are almost equally important (where LE is 51 per cent and H is 49 per cent). At the continental scale (Table 4.1), there is considerable variation in the energy balance recorded by individual continents, which simply relates to the availability (or otherwise) of surface water vapour. For example, Australia is a particularly arid continent, where 69 per cent of the available R_n is used up by H, compared with pluvial South America, where LE dominates the energy balance (64 per cent of R_n).

Table 4.1 *Annual energy balance of the continents (H and LE values are expressed as a % of R_n)*

Area	R_n(%)	LE	H	H/LE(β)	
Europe	100	62	38	0.62	
N. America	100	57	43	0.74	Pluvial
S. America	100	64	36	0.56	
Australia	100	31	69	2.18	
Asia	100	47	53	1.14	Arid
Africa	100	38	62	1.61	

Adapted from Sellers, 1965.

The global and continental energy balances discussed above represent very generalised heat transfers recorded over the long term, when net radiation storage does not occur and energy input equals energy output. However, at the boundary-layer scale, energy balances are controlled by the prevailing site conditions and have a distinctive diurnal–nocturnal rhythm with energy fluxes acting as heat sinks by day (negative signs) and heat sources by night (positive signs). Furthermore, at this microscale, the energy balance is represented by the H and LE fluxes discussed above together with the conduction of heat to (by day) or from (by night) the underlying soil mantle (G), which completes the surface energy balance as follows:

$$R_n = H + LE + G \qquad (12)$$

Box 4.1

The Bowen ratio

1 The Bowen ratio (β) is a useful and simple measure in energy balance representations.

2 It indicates the ratio of sensible heat to latent heat (i.e. H/LE) and reveals the dominance (or otherwise) of an individual flux in an area's energy balance (Table 4.1).

3 A β in excess of 1.00 reveals aridity, with a resistance to vapour flux and a dominance of H. Table 4.1 confirms this association, since Africa, Asia and Australia have β values in excess of this critical value, especially Australia, where it exceeds 2.00.

4 Conversely, a β of less than 1.00 reveals a more humid environment with freely evaporating and transpiring surfaces, and a dominance of LE. This is shown in Table 4.1, where Europe, North America and (especially) South America have β values well below this critical value.

Figure 4.1a illustrates the main energy fluxes involved at any site and represents a daytime situation, when the H, LE and G fluxes act as heat sinks (Figure 4.1b). During the night, when Q_s is zero and I↑ maintains a negative R_n, the three fluxes act as heat sources (Figure 4.1b) for long-wave emissions (see Chapter 3). It must be noted here that other important environmental responses, especially the energy requirements of plant photosynthesis (see Chapter 3, Case Study and Figure 3.5b) and animal respiration, are very insignificant with regard to demands on available R_n (using less than 1 per cent). Figure 4.1a also clearly shows the importance of laterally transported (i.e. advective) energy in the energy balance at a particular site. For example, energy can be advected into (or can diverge out of) the boundary layer in the form of water vapour (LE potential) and sensible heat (H) itself. With advection, the energy requirements at the site can exceed the available R_n and supplementary energy is clearly provided by the advected H. This condition is typical of an oasis situation (Figure 4.2b) found in subtropical desert and polar/tundra environments.

It is apparent that the energy balance within the boundary layer is controlled by the nature of the Earth's surface at the site involved and the relative abilities of the atmosphere and soil to transport heat. The main surface variables concerned are the resistance (or otherwise) to evapotranspiration/vapour flux (see Chapter 6) and the insulation–thermal conductivity properties of the surface materials. These environmental constraints are evident in Table 4.2, which makes qualitative assessments of energy balance differences over a variety of contrasting surfaces. For example, a cornfield or other 'lush' growing crop is dominated by LE (up to 85 per cent of R_n) whereas H is the major heat sink over a dry pavement and desert surface (where LE is almost nil). It should also be noted that leaf litter and a snow pack are excellent insulators and prohibit the G flux, whereas the transparency and turbulent

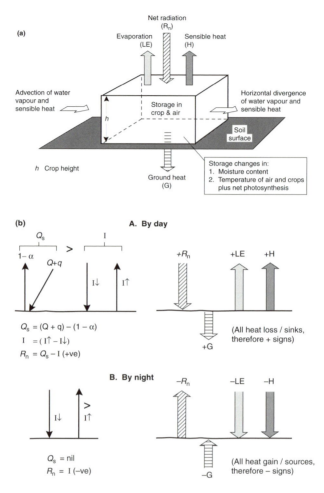

Figure 4.1 *Schematic representations of surface energy balance ((a) after King, 1961).*

mixing of open water are conducive to large amounts of G. However, it must be remembered that no matter which flux process dominates the energy balance over a particular site, it is clear that convection is the principal means of daytime heat transfer away from the surface.

More specific energy balance data are evident in Figure 4.2, which illustrates the diurnal energy balance recorded at three contrasting sites in the USA. The flux patterns were similar at each site, with negative values of net radiation and (most) heat transfer sources before sunrise/after sunset and with distinctive peak positive values (energy sinks) around solar noon. However, the individual R_n partitioning at each site displayed quite considerable variations, mainly associated with the availability (or otherwise) of surface water. For example, Figure 4.2a shows that the Wisconsin grasslands were typical of any mid-latitude vegetated site (see Table 4.2), with R_n dissipated in order of decreasing magnitude by LE, H and G. Furthermore,

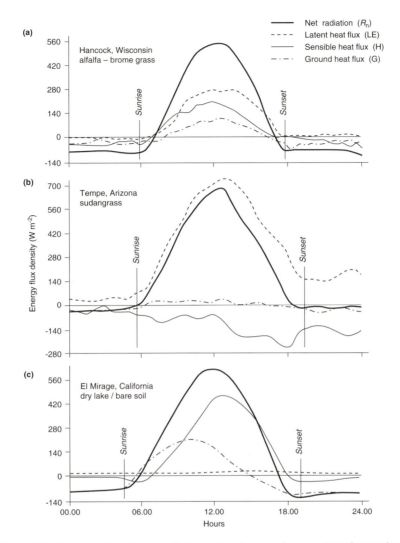

Figure 4.2 *Average diurnal energy balance trends over three contrasting environments (after Sellers, 1965).*

the modest demands made by daytime evapotranspiration (see Chapter 6) were met mainly by the available R_n apart from an hour or so before sunset. This was not the case over the irrigated Sudangrass in Arizona (Figure 4.2b), where LE exceeded R_n at all times and the necessary energy requirements for vapour flux (see Chapter 6) had to be supplemented by H (with a negative β) advected from the surrounding desert (as shown in Figure 4.1a). Finally, at the barren dry lake site in California (Figure 4.2c), evapotranspiration rates and the related LE flux were close to zero, which represented an extreme case of resistance to vapour flux. The major energy sink was now H, although G was dominant between sunrise and 09.30am, when the soil heat capacity reached its maximum.

Table 4.2 *Some qualitative generalisations about the energy balance of a variety of environments*

Type of site	Remarks on the energy balance of the site
Mid-latitude vegetated areas generally	Dissipation modes in order of decreasing magnitude are LE, H and G
Cornfield or other growing crop	LE as much as 80–85% of R_n
Meadow	LE largest dissipation mode during growth; H dominates when the grass is cut, and after it cures and mats to form an insulator, and a vapour barrier
Pavement or stonefield	LE is almost nil; H and G are about equal
Desert	Similar to a stonefield, except H larger than G
Ice cap (high latitude)	LE and G almost nil; H and R_n are about equal and opposite in sign – H positive and R_n negative
Open water	LE and G completely dominate H; LE is at the maximum possible given water and air temperatures; G large because of water's transparency and convective mixing downward below the surface
Still water	G is reduced compared with open water, since convective mixing is reduced. As a result, the water surface becomes warmer, increasing both LE and H
Snow (mid-latitude)	Although snow is transparent to the incident sunlight, its large albedo keeps R_n comparatively small; as long as the snow and air are below freezing, the excellent insulation of snow keeps G small, low vapour pressures keep LE small, and H becomes a large fraction of R_n. Snow at the freezing temperature results in melting and mass transfer within the G term
Leaf litter	The litter acts as a thermal insulator, keeping G very small; acts as a vapour barrier when dry; H is largest mode of dissipation

After Lowry, 1967.

CASE STUDY 4: Radiation balance and energy balance modifications due to urbanisation

In the case study at the end of Chapter 3, it was clear that inadvertent human activities have the potential to seriously alter natural global radiation balances through the enhanced greenhouse effect and global warming. In this study, it will be become apparent that urbanisation can play a significant role at the more local boundary-layer scale, with important changes in net radiation and energy balance partitioning. Research into the impact of human activities on boundary-layer radiation balance and heat transfers has been neglected in the past. Instead, attention has been focused on the more global (and popular) issues of acid rain, ozone depletion and greenhouse warming (as discussed in the three case studies so far). Indeed, little progress has been made in understanding energy balance changes in the boundary layer, apart from the urban environment, which will be discussed below.

Net radiation (i.e. the all-wave radiation balance, R_n, discussed in Chapter 4 and expressed in equation (10)) has been affected by urbanisation.

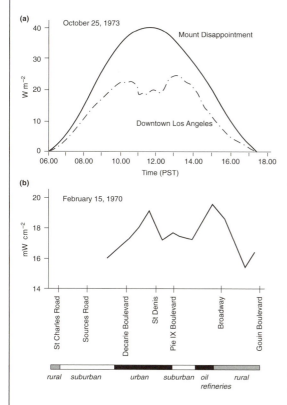

(a)

October 25, 1973

Mount Disappointment

Downtown Los Angeles

Time (PST)

(b)

February 15, 1970

St Charles Road | Sources Road | Decarie Boulevard | St Denis | Pie IX Boulevard | Broadway | Gouin Boulevard

rural   suburban   urban   suburban oil   rural
                                         refineries

Figure 4.3 *(a) Ultraviolet short-wave radiation variations over Los Angeles; (b) incoming long-wave radiation variations over Montreal ((a) after Petersen and Flowers, 1977; (b) after Oke, 1987).*

The limited data available clearly reveal that the urban fabrics and structures have modified each one of the flux components concerned (namely: $Q + q$, or incoming short-wave radiation, SW↓; $1 + \alpha$ or outgoing short-wave radiation, SW↑; I↓ incoming long-wave radiation, LW↓ and I↑ outgoing long-wave radiation, LW↑). For example, SW↓ measured over Montreal (Canada) and Newcastle (Australia) revealed depletion rates of between 2 and 10 per cent, due to the increased dust-veil scattering in the high aerosol concentrations over urban areas (see Chapter I and Table 1.3). This depletion was confirmed by measurements from Los Angeles (USA), when, on a polluted day, ultraviolet SW↓ radiation received 'downtown' was 53 per cent less (around solar noon) compared with the rural Mount Disappointment (Figure 4.3a). However, despite these important losses, the urban/rural short-wave radiation balance differences are small since outgoing SW↑ is reduced due to the reduced albedo (typically 5–10 per cent lower) of the urban fabric.

Similarly, measurements of the long-wave (infrared) radiation balance show quite large changes in the individual LW↑/LW↓ fluxes over the urban area, which are counterbalanced again to produce small overall changes. For example, the urban heat island has high rates of emissivity (see equation (8)) and outgoing LW↑ radiation. However, this loss is counterbalanced by large amounts of incoming LW↓ radiation (averaging 11 per cent more over Hamilton, Canada) associated with the enhanced greenhouse effect (see Chapter 3, Case Study) in polluted urban/industrial air (with increased CO_2 and water vapour concentrations). Figure 4.3b illustrates LW↓ trends over Montreal (Canada), where counter-radiation peaks with the polluted urban air and with the greater water vapour concentrations over the city's oil refineries. Because of all these short-wave and long-wave counterbalances, overall net radiation (R_n) differences are small between urban and rural areas (generally less than 5 per cent). For example, in Newcastle (Australia), the summer R_n differences averaged +2.9 per cent (i.e. rural areas are higher), whereas the winter differences averaged −5.5 per cent (i.e. urban areas are higher, due to greater LW↓ in the polluted air).

The urban sprawl modifies energy balance by changing the surface albedo, replacing vegetation with the urban fabric (increasing vapour flux resistance) and by emitting vast amounts of sensible heat and aerosols. The urban climatological literature suggests that the city is a desert or karst environment with Bowen ratio values (see Box 4.1) well in excess of 1.0, sensible heat flux the dominant energy sink and decreased atmospheric humidity. Traditional urban climatic data are related to the last parameter and it appears that relative humidity values in the city (compared with rural areas) range from 2 per cent less in winter to 8 per cent less in summer, which represents the peak heat island season. Energy balance data are limited for urban areas and the most pertinent available data have been collected for Montreal, Los Angeles and Vancouver in North America (Table 4.3). Reference to Table 4.3 reveals interesting heat transfer–city relationships. Indeed, Los Angeles appears to be a 'textbook' dry city,

Table 4.3 *Heat transfers over urban areas*

City	Bowen ratio (β)	H	LE (% of net radiation)	G
Montreal	0.75	20	70	10
Los Angeles	>1.00	80	0	20
Vancouver	>1.00	51	35	14
		67[1]	20	13
		37[2]	47	16
		49[3]	38	18

Notes: [1] Dry, 10 days prior without rainfall
[2] Wet, immediately after a record 8 cm rainfall
[3] Drying out, four days after rainfall

Source: After Yap and Oke, 1974; Terjung *et al.*, 1970; Oke *et al.*, 1972.

since the Bowen ratio value exceeds 1.0 and is quoted as approaching infinity, with zero vapour flux.

This was attributed to a long summer drought in southern California, where fabric storage of water was at a minimum. Furthermore, the supply of water vapour by combustion (related to automobile, industrial and domestic sources) was also negligible. This latter point appears rather anomalous in Los Angeles, a city with a notorious photochemical smog, with a potential for appreciable amounts of exhaust water vapour. The Montreal data are completely different and reveal that this city is not arid or karstic. It is freely evaporating and 'wet' with a mean Bowen ratio of 0.75 and with latent heat flux consuming 70 per cent of the net radiation. The source of unexpected water vapour appears to be related to the *in situ* combustion sources, especially automobiles and oil refineries in the urban area. It was also assumed that the urban fabric stores more water than was hitherto thought.

Figure 4.4 illustrates the diurnal variation of Bowen ratio for Montreal *vis-à-vis* β values for typical rural vegetated areas, namely alfalfa-brome grass in Wisconsin and Douglas fir forest in British Columbia. The diurnal course of the Bowen ratio in the city is unique. Unlike rural surfaces, the city does not show negative β values at night, due to the heat island effect, which makes inversions (see Case Study 5) rare in the urban boundary layer. Conversely, nocturnal negative values characterise the rural sites (remote from the heat island), associated with temperature inversions when the energy for vapour flux was supplemented by sensible heat transfers. Apart from a β peak over Montreal around solar noon, (β reached 1.6, with H temporarily dominant at this time of excess net radiation), it appears that the greater proportion of energy is used for latent heat transfers. The 'unexpected' vapour source is explained above in terms of the combustion of fossil fuel.

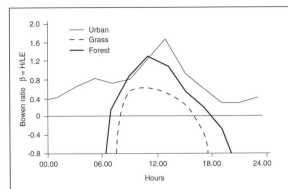

Figure 4.4 *Bowen ratio variations over three contrasting environments (after Oke et al., 1972).*

Comparison of urban patterns and forest/grassland values will emphasise the significance of human interference (heat island development) with urbanisation. The forest also had peak β values around 1.2 at solar noon (H temporarily exceeded LE, as over the city) but at all other times was freely evaporating with strong nocturnal inversions. The grass β values did not exceed 0.5, to indicate that the vapour flux was greatest over this surface and indeed negative values were dominant for most of the period.

The Vancouver data confirm an earlier supposition that a considerable amount of water is stored in the urban fabric and is available from combustion sources. Consequently, latent heat flux consumed 20 per cent of the net radiation during 'dry' conditions, following a ten-day period without rainfall. During the 'wet' conditions of observation (2), the latent heat flux was temporarily the dominant energy sink. However, the sensible heat dominance returned within four days of the record rainfall total. It is apparent that a five-day drying period (following substantial rainfall) was required for the city to recover the original (pre-rainfall) moisture state with a resistance to vapour flux, as in observation (1). It appears that Vancouver compares favourably with the Los Angeles heat transfer pattern (H > LE), although vapour flux was always experienced in the former city, suggesting the importance of the combustion source. For a day or so during the wet period, Vancouver had energy balance characteristics similar to those in Montreal (i.e. LE > H) even though the magnitude of the vapour flux dominance was smaller in the west coast city. The variability of the Vancouver transfers over the ten-day period represented by the observations emphasises the significance of long-term synoptic fluctuations in the assessment of heat transfers over all environments.

Energy transfers within the troposphere and their influence on atmospheric stability tendencies

Energy transfers take place within the troposphere in free and forced convective systems, when parcels of air rise through the atmospheric environment, with separate temperature changes. This relationship controls the amount of vertical movement within the troposphere, since it leads to stable or unstable conditions. This chapter covers:

- the dynamics of free and forced convection
- the temperature changes within the atmospheric environment (the ELR)
- temperature changes in dry (DALR) and saturated (SALR) parcels of air
- ELR and DALR/SALR relationships: stability and instability
- case study: temperature inversions and their role in pollution episodes

The previous chapter has emphasised the distribution of net radiation and the transfer of energy at the Earth's surface on the global and local scales. Without these transfers, the atmosphere would be heated to an intolerable degree within the boundary layer, which would accentuate the latitudinal imbalance of insolation observed between the equator and the poles (see Figure 3.3). It should be noted that advective transfers of sensible and latent heat also help to offset the excessive surface heat accumulation in the tropics associated with the movement of energy in ocean currents (e.g. the Gulf Stream /North Atlantic Drift; see Figure 12.2) and airstreams, such as the prevailing south-westerly winds in mid-latitudes. It is apparent that subtropical ocean currents alone account for about 25 per cent of the energy transported out of low attitudes.

Free and forced convection

This chapter examines the transfer of energy within the troposphere by free and forced convective processes, as discussed in Chapter 4, and considers the temperature changes that occur in these rising air parcels *vis-à-vis* the surrounding atmospheric environment. This leads on to a discussion of the thermal and density relationships between rising parcels and the surrounding air in terms of basic stability tendencies, which control the degree of free convection in particular.

Free convection refers to air parcels that are heated initially through contact with surface heat accumulation (or lapse rate steepening, as explained later), associated mainly with local insolation irregularities (i.e. due to albedo differences; see Table 3.2). For example, the lower albedo of a city's bitumen (0.02) contrasts markedly with that over rural grasslands (0.20). This partly accounts for the urban heat island, which results in the greater free convection, cloud cover and precipitation over urban areas (see Chapter 16). When the air parcel becomes superheated, it develops a higher temperature and lower density (with expansion) compared with the surrounding air. The warmer, less dense and more buoyant air parcel now rises from the surface (like a hot air balloon vigorously heated by gas burners) and is replaced by cooler, more dense air converging towards the area of excessive heating. The more buoyant air parcel will continue to rise as long as it remains warmer (and less dense) than the surrounding atmospheric environment. If the temperatures/densities become the same, equilibrium is achieved (see Box 5.1) and the air will diverge horizontally. Finally, if the parcel becomes colder and denser than the atmospheric environment, then it is forced to descend through the atmosphere as an area of pronounced subsidence and surface divergence (as in the 'eye' of a hurricane, discussed in Chapter 15).

Forced or mechanical convection represents the vertical transfer of energy by eddy currents in the atmosphere, which are mainly associated with obstruction to smooth or laminar air movement (see Figure 16.5a). It also takes place along active fronts (e.g. the Polar Front in Chapter 12), which represent boundaries of air masses with contrasting temperatures and densities. Strong orographic and frontal displacement promotes a vigorous, steady updraught of air and is responsible for mixing air with contrasting temperature, humidity and density from upper and lower levels. The resultant circulations can lead to distinctive cyclogenesis and the formation of orographic lows/lee-wave depressions (e.g. the Genoa Low over north-west Italy) and wave depressions in middle latitudes (e.g. the infamous British Storm of October 1987), which are discussed in Chapter 14.

As mentioned above, air parcels forced to rise by these convective processes can transfer energy (and indeed mass) to great heights only if they can remain warmer, less dense and more buoyant than the surrounding air (*viz.* the atmospheric environment). This important relationship is indeed vital for these transfers and is controlled by the basic prevailing temperature and density contrasts experienced between the atmospheric environment and the rising parcels of air, which are rarely in a state of equilibrium (see Box 5.1).

Temperatures within the atmospheric environment

Chapter 2 briefly referred to the temperature distribution in a large mass of stationary troposphere, which represents the lowest part of the atmosphere, where the atmospheric environment/air parcel differences are most effective in influencing weather processes/systems. It was noted that the troposphere normally cools with increasing height at an average environmental lapse rate (ELR) of 6.5 °C 1000 m^{-1}. It was also noted that the ELR was indeed highly variable, with the 'average' rate distorted by extreme variations associated with temperature inversions and lapse rate steepening (explained below).

It must be reiterated here that the ELR refers to temperature conditions existing in a large mass of stationary air (in vertical terms) at a given place and time. It must be remembered that this air has no vertical movement and represents the total atmospheric environment through which parcels of air are moved by free and forced convection. Indeed the ELR is approximately 1000 times greater than the average horizontal rate of temperature change with latitude. Also, the 'normal' cooling rate is mainly associated with increasing elevation and with increasing remoteness from the Earth's surface, which is the source of air heating by radiative and weak conductive processes. Other factors affecting the cooling rate include the decreasing amount of water vapour recorded aloft, which leads to a weak greenhouse effect (see Case Study 3) and a marked reduction in the absorption of infrared radiation emitted from the Earth's surface. Finally, the air becomes progressively less dense aloft, and the 'thinner' air means a greatly reduced heat retention and absorption of terrestrial radiation.

It has been noted that the average ELR is represented by a cooling with increased elevation in the order of 6.5 °C 1000 m^{-1}. However, there is nothing constant about this rate of change, and in fact the ELR is highly variable due to dramatic short-term changes in the receipt of insolation at the Earth's surface. Also, the concentration of water vapour and heat in the troposphere can fluctuate dramatically with advective/airstream changes. These ELR variations are clearly shown in Figure 5.1a, ranging from a constant decrease with height (condition A) to a pronounced increase (condition C) or decrease (condition B, which is more or less isothermal). However, the most extreme deviations from the average cooling trend are associated with lapse rate steepening (condition E) and a temperature inversion (condition D). Indeed conditions D and E represent the typical diurnal variation of the ELR under clear skies, following terrestrial radiation loss by night (leading to inversion D) and solar radiation accumulation (with steepened lapse rate E) by day.

Lapse rate steepening (or super adiabatic conditions described later) is characterised by a very rapid temperature decrease (approaching 14

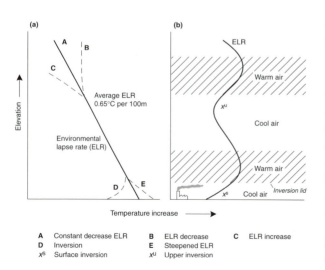

A Constant decrease ELR   B ELR decrease   C ELR increase
D Inversion   E Steepened ELR
X^S Surface inversion X^U Upper inversion

Figure 5.1 Lapse rates (ELR) in the troposphere showing (a) varying rates and (b) surface and upper-air inversions.

°C 1000 m^{-1}, or more than twice the average ELR) in the layers of air close to a very hot Earth's surface. It develops on hot, sunny, summer days with excessive heat accumulation in the 'skin' of air next to the surface. As will be seen later, these conditions can develop into extreme instability and the formation of severe thunderstorms (see Chapter 8). However, the most dramatic ELR deviations are associated with temperature inversions, which reverse the normal (cooling) lapse rate trend when (occasionally) temperature actually increases with height (condition D in Figure 5.1a). These ELR reversals are clearly illustrated in Figure 5.1b, with inversions recorded at surface layers (X^s) and aloft (X^u), which develop in markedly different ways. The origin of both surface and upper-air inversions will be explained in the case study at the end of this chapter, which emphasises the respective roles of nocturnal terrestrial radiation loss (on long, clear nights) and subsidence warming in strong anticyclonic divergence.

Adiabatic temperature changes in parcels of air ascending/descending through the troposphere

So far in this chapter, we have discussed the variable characteristics of the atmospheric environmental lapse rate through which parcels of air are forced to rise by convective processes. Furthermore, when a parcel is moved upwards from the Earth's surface, it is subjected to very important temperature and density changes during its ascent through the surrounding atmospheric environment. As the parcel rises and moves into a region of lower pressure, it expands and cools, since it uses up internal potential heat to supply the energy for the expansion process (analogous to a gas-filled balloon that bursts with height).

Conversely, descending air parcels will encounter increasing air pressure and will experience compressional warming (analogous to the warming air in a tyre when pumping takes place). It must be noted that both these cooling and warming processes are isolated from the surrounding atmospheric environment, since air is a poor conductor of heat. Consequently, no energy enters or leaves the parcel from/to the outside air, and expansion cooling/compressional warming take place at an independent, self-contained rate, which is termed 'adiabatic'. Essentially, all large-scale vertical motion in the troposphere involves these adiabatic changes. Also, all subsiding air parcels will always warm at the same rate at which rising air parcels cool (while ascending to an equivalent height without saturation, see Chapters 6 and 7). Furthermore, since no energy crosses the parcel boundaries, it recovers its initial state conditions upon returning to its original elevation. This rate of temperature change is constant for dry, unsaturated air and is known as the dry adiabatic lapse rate (DALR), which is fixed at 10 °C 1000 m^{-1} (compared with the variable 6.5 °C 1000 m^{-1} ELR discussed above).

However, rising parcels of air cool by expansion and soon reach dew point (see Chapter 7), when condensation occurs and the air becomes saturated. The rising air

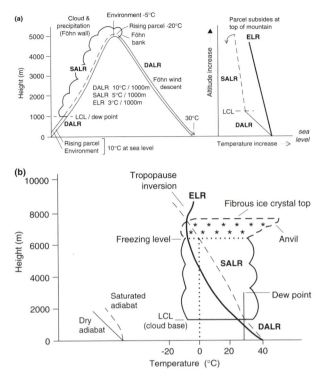

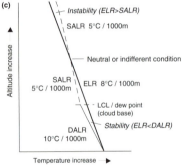

Figure 5.2 *Schematic representations of equilibrium conditions in the troposphere: (a) absolute stability; (b) absolute instability; and (c) conditional instability.*

parcel will still cool by expansion, but once saturated it will release latent heat of condensation (2.45 MJ kg⁻¹ at 20 °C; see Chapter 4), which must slow down the rate of cooling within the parcel (caused by expansion) by 5 °C 1000 m⁻¹ on average. The reduced rate of cooling following dew point attainment is known as the saturated adiabatic lapse rate (SALR). It is most variable (unlike the fixed DALR) since it depends on the amount of water vapour present within the parcel and the temperature, which controls the degree of parcel saturation. As a result, the amount of liberated latent heat is much greater at high temperatures, so that SALR cooling is slower in 40 °C air compared with 0 °C air. Conversely, at temperatures below –40 °C (found in polar regions and very high elevations), the SALR approximates to the DALR, since only minute amounts of water vapour are being converted into ice crystals by sublimation (see Chapter 7), and only negligible amounts of latent heat are liberated. A 'rule of thumb' is that for temperatures above 30 °C, the SALR is 5 °C 1000 m⁻¹, while below 0 °C, the rate increases to 7 °C 1000 m⁻¹ and reaches 10 °C 1000 m⁻¹, which is the DALR, at –40 °C.

When moist air subsides, the resultant adiabatic warming must follow either the dry or saturated rates. Without evaporation, the DALR operates for the descent, whereas with evaporation, latent heat is used up to effect the change from liquid to vapour and so warming proceeds at the SALR. This principle is best explained with reference to the föhn/foehn/Chinook wind (see Chapter 16), which is illustrated in Figure 5.2(a). The parcel ascent on windward slopes takes place mostly (i.e. above dew point at 1000 m) at the SALR, with a net gain of 20 °C from the release of latent heat (i.e. at 5000 m, the parcel is now –20 °C compared with –40 °C if the DALR had operated up to this summit). The descent on the leeward slope is dominated by the DALR (with evaporation and the SALR confined to maintaining the föhn bank; Figure 5.2a). This effect transports most of the 20 °C net heating to the foot of the leeward slopes as very warm (30 °C), dry wind, which accelerates the ablation of the winter snow pack. This process is termed 'pseudo-adiabatic', since there is an increase in the parcel's potential temperature if it is brought adiabatically to a standard pressure of 100 kPa. It is a relatively rare phenomenon, only occurring to the lee of the highest mountain ranges in the Rockies, Alps or Andes under stable equilibrium (see next section and Figure 5.2a).

The relationship between the ELR and DALR/SALR: stable and unstable atmospheric conditions

---

Box 5.1

Atmospheric equilibrium

1 Air parcels are given vertical motion by free and forced convective processes, due to lapse rate steepening and mechanical uplift mechanisms discussed at the start of this chapter.

2 These parcels ascend and descend through the atmospheric environment with distinctive (if self-contained) adiabatic temperature changes.

3 These changes are represented by the DALR in dry air and the SALR in air that has reached the condensation/dew-point level.

4. However, the rate and amount of vertical motion that can develop depends mainly on the type of thermal balance/equilibrium that exists within the troposphere.

5 Furthermore, the nature of this balance is determined by the prevailing thermal/density conditions, particularly the relationship between ELR and the DALR/SALR, which is a vital consideration for everyday weather forecasting.

6. When this relationship results in the loss of buoyancy of the air parcel, it must come to rest and eventually sink downward in a condition of stability.

7 Conversely, if uplift results in increasing buoyancy, then the parcel can move further away from the source of displacement up to considerable heights (and towards the tropopause, as indicated in Figure 1.1) in a condition of instability.

Stability occurs when the current ELR is less than the DALR (Figure 5.2a), which means that the rising air parcel will initially cool much faster than the surrounding atmospheric environment. As it rises, the parcel becomes colder and heavier than the air around it and the dense parcel will sink back to its former position. Indeed, such an air parcel only rises in the first place when a forcing process (normally a mountain or front) initiates the displacement, since free convection is prohibited by the unfavourable density relationships at ground level. However, if equilibrium is restored just above dew-point level, then free convection is allowed over a very limited vertical distance. Under these conditions, the ascent is quickly terminated and the resultant clouds can only develop into very shallow (fair weather) cumulus humilis with distinctive flattened tops (see Figure 8.1 and Plate 8.3), which do not allow precipitation (see Chapter 8).

When the ELR is lower than both the DALR and the SALR, then absolute stability exists, which restricts upward movement of the air parcel even when condensation and latent heat release occur. This is the condition represented by Figure 5.2a when non-buoyant, dense air is forced over a 5000 m mountain range. Despite the continued orographic uplift of dense and stable air parcels, at the summit the parcel is 15 °C colder than the surrounding atmospheric environment and the greater density forces the parcel to sink down the leeward slope (and warm by compression at the DALR to become the föhn wind, discussed earlier). The cloud cover above the lifting condensation level (LCL) or dew-point level on the windward slopes is shallow stratus or stratocumulus, which tends to literally 'hug' the slopes and mirror the topography involved (Plate 5.1). Conversely, to the lee of the crest below the distinctive föhn bank (Plate 5.2), the cloud quickly dissipates to produce sunny, dry and increasingly warmer slopes. With forced frontal displacement, stability restricts cumuliform cloud (see Figure 14.6) development, leading to kata frontal conditions in a wave depression (see Chapter 13), with layered cloud/drizzle at the warm front and stratocumulus/light rain at the cold front.

Plate 5.1 *Absolute stability – clouds 'hugging' the active volcano Mount Ruapehu, New Zealand.*

Plate 5.2 *Föhn bank over Mount Sefton, New Zealand.*

Absolute instability occurs (Figure 5.2b) when the prevailing ELR is greater than both the DALR and the SALR in the lower and middle troposphere. Consequently, the parcel will cool more slowly during its ascent and rapidly becomes less dense and more buoyant than the surrounding atmospheric environment. This occurs particularly with lapse rate steepening/super-adiabatic conditions (discussed earlier), when the increasing buoyancy exaggerates the vertical motion and the rising air parcel accelerates freely up to the level (i.e. the tropopause) where its temperature eventually reaches that of the surrounding air. This is illustrated in Figure 5.2b where free convection occurs and both the DALR and the

Plate 5.3 *Absolute instability – cumulus congestus (viewed from Point Lookout) over the New South Wales Tablelands, Australia.*

SALR are lower than the ELR. At the LCL, the parcel is already 4 °C warmer than the surrounding air, and this difference increases as the SALR takes over (which is about one-half the ELR). Consequently, at 3000 m, the parcel is 10 °C warmer (and more buoyant/unstable) than the air around it. It continues to rise freely until about 7500 m, forming cumulus congestus clouds (Plate 5.3) of great vertical thickness, until the tropopause is reached and the ELR decreases markedly in the upper inversion/isothermal trend at this altitude (Figure 5.2b). Chapters 8 and 16 will reveal that deep cumiliform clouds form in such unstable conditions with torrential rain and thunderstorms. Furthermore, these cumiliform clouds and heavy rain characterise both warm and cold fronts in a wave depression, which are termed ana with the rapid uplift of buoyant and unstable air (see Case Study 14).

Conditional instability is a more common occurrence in mid-latitudes and happens when the ELR lies between the DALR and the SALR throughout the entire ascent of the parcel. This situation is illustrated in Figure 5.2c and it is evident that in the early part of the ascent, when the air is unsaturated (below the LCL), the DALR exceeds the ELR (*viz.* 10 °C 1000 m^{-1} compared with 8 °C 1000 m^{-1}). This makes the parcel colder and heavier than the surrounding air and, at this level, the parcel becomes denser and resists displacement. Consequently, any further rise can take place only with forced convection, especially when a parcel is forced over an active cold front (see Chapter 14) or a high mountain range.

However, such forced uplift will eventually move the parcel up to the LCL, where latent heat is released and the SALR, which is lower than the ELR (e.g. in Figure 5.2c

it is 5 °C 1000 m^{-1}, compared with an ELR of 8 °C 1000 m^{-1}), takes over. From now on, the atmospheric environment is cooling more quickly than the parcel and ultimately, beyond the neutral/indifferent condition (Figure 5.2c), the rising parcel becomes increasingly warmer than the surrounding air, with a dramatic change to buoyancy/instability. The parcel now rises freely on its own accord and the cloud 'mushrooms' into a cumulus congestus form of great vertical extent (Plate 5.4), with torrential rain showers. Obviously, this dramatic change in

Plate 5.4 *Conditional instability – cumulus congestus clouds over the highest peaks in the Colorado Rockies, USA.*

equilibrium from stability to instability (and the associated extreme weather phenomena) is conditional upon the release of latent heat of condensation, which reduces the adiabatic cooling to a rate much less than the ELR.

CASE STUDY 5: Temperature inversions and their role in pollution episodes

It is clear from discussions so far that occasional temperature inversions at the Earth's surface and in the upper air are important reversals of the normal ELR in the troposphere (Figure 5.1a, condition D and Figure 5.1b, X^s and X^u). Ideal conditions for surface inversions are those that are conducive to rapid cooling of the Earth's surface and adjacent air by nocturnal, infrared terrestrial radiation leaving warmer air at higher elevations. For example, a long winter night is highly effective in maximising the loss of long-wave infrared radiation into the upper atmosphere and (through the 'window') into space (see Case Study 3). Furthermore, this loss is accentuated by clear skies and dry air (with weak counter-radiation and greenhouse warming) and calm conditions. These allow the air to stagnate and be chilled slowly by conduction from below (remembering that air is such a good insulator).

Obviously, the best synoptic conditions are associated with 'blocking' winter anticyclones (see Chapter 9, and Case Studies 11 and 12) over a snow-covered surface, particularly in areas of diverse relief. Then, cold-air drainage or katabatic airflow (see Chapters 7 and 16) into valleys intensifies the radiative chilling, which leads to quite severe (if localised) inversions. The classic example of this occurred in a 100 m deep Gstettneralm sink hole near Lunz, Austria, on the night of 21 January 1930, when the temperature at the base of the depression was recorded as −28.8 °C, compared with +2.3 °C at the surface. Besides being produced by in situ radiative chilling, surface inversions are also caused by the advection of warm airstreams over a cold surface (land or

water) that is chilled from below by conduction. Perhaps the best example occurs in central California, in the San Francisco Bay area, when cool maritime air from the offshore cold current wedges under warmer continental air. This causes a strong surface inversion with widespread advection/ marine fog (see Chapter 7 and Plate 7.2).

Upper-air inversions can develop at various levels in the troposphere and are mainly associated with persistent (i.e. over months) subsidence warming in strong anticyclones (see Chapters 9 and 13). In the subtropics, semi-permanent anticyclones on the poleward limb of the Hadley cell are widespread, due to jet stream convergence aloft, and are associated with dynamic subsidence over 15,000 m, which feeds strong, surface divergence and outflowing trade winds (see Chapter 9). During this prolonged dynamic descent, the air encounters increasing pressure and is heated by compression (see Chapter 5) so that a layer of warm, dry air accumulates around 2000 m elevation, towards the base of the subsidence. It should be noted that this warming does not extend to the Earth's surface, since nocturnal and low-sun seasonal chilling (at a desert surface in particular) counteracts the heat accumulation in the lowest air layers. Furthermore friction, associated with land-surface features, creates strong eddy currents, which clearly dissipate any heat build-up, so a situation develops with a layer of warm air accumulating aloft above cooler surface air, which represents an upper inversion (i.e. X^u in Figure 5.1b and Figure 5.3, condition B).

In the subtropics, this heat accumulation in the

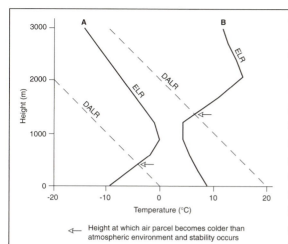

Figure 5.3 *Stability in dry air due to the occurrence of (a) a surface inversion and (b) an upper-air inversion.*

upper air is termed the 'trade wind inversion', which is especially well-developed and lowest (1000–2000 m) over the eastern side of the oceanic trades (e.g. in the Azores region of the central Atlantic). However, it must be noted here that the subsidence inversions are regularly found outside the subtropics, since they dominate the middle to upper air of all deep, persistent anticyclones, including the so-called 'blocking highs' (see Case Study 11, and Chapter 13) that regularly influence mid-latitudes (including north-west Europe). Upper-air inversions also occur with warm front advection, associated temporarily (i.e. less than a day) with advancing wave depressions in mid-latitudes (see Chapter 14). This type develops ahead of the warm front (see Figure 14.5a) when less dense, warm and moist (tropical maritime) air overrides a 'wedge' of cold (polar continental) air ahead of it. About 300 km ahead of the front, the warm air can be forced up to 7000 m and represents an area of marked upper tropospheric warming compared with the colder air below this advection zone.

Surface and upper-air inversions have an important role to play in the concentration of pollutants in the troposphere, since they are responsible for strong atmospheric stability (see previous section), which tends to inhibit vertical motion and aerosol dispersal. Because of the presence of a surface inversion (Figure 5.3, condition A), the DALR intersects the ELR at a much lower elevation that marks the height at which the rising air (and aerosol dispersal) will come to a halt and eventually sink back to Earth, leading to fumigation-type pollution. Severe pollution episodes develop that can lead to human discomfort and an increased death rate (e.g. the infamous London smog of December 1952, which killed 4000 people and is discussed in Chapter 7). Ironically, pollution control measures that followed this disaster recommended the construction of chimneys and stacks well above the inversion lid (i.e. into the more buoyant air). This dispersed the pollutants more readily to avoid *in situ* problems, but long-range transport in the westerly airflow soon led to the acid rain problem across the North Sea, in Sweden and Norway in particular (see Case Study 1).

Likewise, upper-air inversions (Figures 5.1b and 5.3 condition B) are responsible for strong atmospheric stability aloft, which also inhibits the formation of deep turbulent eddies and the removal of pollutants. The concentration of aerosols in the lowest 1500 m of the troposphere (i.e. trapping) causes very serious pollution episodes over subtropical cities like Los Angeles, Johannesburg, Mexico City and Sydney. This is especially so since the abundant sunshine in these latitudes introduces photochemical reactions with pollutants (especially hydrocarbons and oxides of nitrogen) to produce the notorious photochemical smog (including ozone and peroxyacetyl nitrate or PAN). More temperate cities also experience similar pollution problems with summer blocking highs, when residents in Athens, London and Tokyo suffer eye–nose irritation, headaches and chronic lung diseases, which were first recorded in the Los Angeles region in the 1930s.

Key topics for Part II

The atmospheric radiative fluxes and energy balances initiate and sustain a wide range of atmospheric processes and interactive systems:

1   Solar radiation received at the Earth's surface (termed insolation) is represented by a large flux and short wavelength. It is depleted during its passage through the atmosphere by absorption, reflection and scattering.

2   The supply of solar radiation is balanced by energy-dispersal mechanisms, including long-wave/small-flux terrestrial radiation and sensible/latent energy transfers. The radiative losses are controlled by the effectiveness of the greenhouse effect, when enhancement accentuates the absorption and re-radiation of infrared radiation back to the Earth's surface, promoting global warming.

3   The balance between incoming/outgoing solar radiation and incoming/outgoing terrestrial radiation is termed net radiation, which varies spatially and temporally.

4   Net radiation is totally partitioned between sensible and latent energy transfers, which again vary spatially and temporally at every scale (global to micro), representing the energy balance at the Earth's surface.

5   Energy is transferred vertically within the troposphere by free and forced convective systems, in the form of sensible and latent heat. However, this transfer to great heights is possible only if the rising air parcels (cooling adiabatically) can remain warmer than the surrounding atmospheric environment. This equilibrium condition is termed instability and leads to stormy weather from towering cumulonimbus clouds.

6   Due to nocturnal surface chilling and subsidence warming aloft, temperature inversions can develop within the troposphere and lead to serious air pollution episodes as they accentuate stability within the air mass.

Further reading for Part II

Global Air Pollution. Howard Bridgman. 1990. Belhaven Press.
Provides a clear and detailed study of the role of trace gases in the enhanced greenhouse effect.

Atmospheric Pollution. Derek Elsom. 1987. Blackwell.
A comprehensive and clear examination of the atmospheric pollution problem, covering both the nature, sources and effects of pollution and national/international approaches to pollution control.

Climate Change 1995: The Science of Climate Change. J. T. Houghton *et al.* (eds) 1996. Cambridge University Press.
The latest, definitive statement by a panel of experts (the IPCC) on the predicted magnitude and environmental consequences of global warming.

Global Environmental Issues. David Kemp. 1994. Routledge.
A concise and non-technical account of the greenhouse effect and global warming.

Weather and Life. William Lowry. 1967. Academic Press.
Over 30 years old, but this still remains a concise and clear account of the concepts and applications of radiation, energy balances and the diffusion of air pollution.

Boundary Layer Climates. T. R. Oke. 1987. Routledge.
Remains the most up-to-date, concise and non-mathematical approach to radiation and energy processes in a wide range of natural and person-modified atmospheric environments.

Part III Atmospheric Water

Parts I and II confirmed that the atmosphere contains a significant if variable amount of water vapour, which has a vital role in the operation of atmospheric processes and systems at every scale. This contribution varies from the global scale (through the efficiency of water vapour in the greenhouse effect; Chapter 3) to the more local scale, with the liberation of the latent heat of condensation in rising air parcels (see Chapter 5). This is responsible for föhn wind warming (see Chapter 5) and the intensification of cyclones and depressions in the troposphere (see Chapters 13, 14 and 15).

Part III is concerned with water vapour in terms of its potential for initiating and sustaining the pathways in the hydrological cycle (see Figure 6.1). This represents the continuous passage of water from a liquid state at the Earth's surface to a gaseous state within the atmosphere (through evaporation/transpiration) and its return to the surface in a liquid or solid form (after condensation/precipitation). These are the component parts of the cycle (known as pathways) as far as the atmosphere is concerned and are explained in detail in the following three chapters.

6 ▸ Evaporation and evapotranspiration

No discussion of atmospheric water would be complete without a consideration of the processes and pathways by which air initially acquires its water content. This chapter covers:

- definitions of terms used
- explanation of processes involved
- discussion of the roles of meteorological and environmental factors
- case study: human suppression of vapour flux

Box 6.1

Vapour flux processes and terminology

1. Evaporation is simply defined as the process by which liquid or ice is converted into a gas, which is transferred into the troposphere by convective processes (see Chapter 5).

2. This vapour flux from the Earth's surface is primarily from the oceans, particularly in low latitudes, where solar radiation surpluses (see Figure 3.3) promote excessive amounts of evaporation (e.g. 200 cm yr^{-1} in the western Pacific and Indian Oceans, at latitude 15° S).

3. Evaporation also takes place from terrestrial water bodies (ranging in size from large lakes to small ponds or even puddles) and exposed wet soil.

4. The flux from snow and ice surfaces is known as sublimation, which takes place directly from a frozen state to vapour, bypassing the liquid stage.

5. Another important vapour flux is the transpiration of moisture through the pores and stomata of plants, which is partly acquired from the adjacent soil water in the root zone.

6. Consequently, this combined plant/soil flux is more correctly termed evapotranspiration and, as is discussed later in this chapter, it is further subdivided into actual and potential rates.

Evaporation and transpiration processes

It should be noted that water molecules are in constant motion, whether in a large water body or a thin film covering soil or leaves. Furthermore, when heat is added to the water, the molecules become increasingly energised or excited and move more rapidly. Eventually, the molecules possess sufficient energy to break through the restrictive membrane of the water surface and diffuse into the atmosphere in a gaseous form. Similarly, some of the water molecules contained in this water vapour in the lowest atmosphere, which are also in motion, may penetrate the water surface and enter the liquid. Therefore, the rate of evaporation at a given time will depend on the number of 'fast' molecules leaving the water surface less the number of returning molecules. A positive exchange (i.e. more molecules leaving the water surface than are returning to it) will result in evaporation, whereas the reverse exchange will terminate vapour flux and initiate condensation (see Chapter 7).

The molecules making up a volume of liquid water are in close proximity, with a separation of just over one molecular diameter. At such distances, the molecules interact with a strong bonding or attraction, and a short-range force develops that weakens with increased separation. In water vapour, the molecules are much further apart (typically ten or more molecular diameters), although the distance involved depends on the actual weight of the water molecules (or vapour pressure). Furthermore, with this increased separation, the intermolecular bonding (or molecule interaction) is very small indeed. Therefore, in order to create water vapour from liquid water, it is necessary to increase the separation between all the molecules by supplying kinetic energy (from solar radiation via sensible heat transfers) to work against the holding force. The amount of energy required is directly related to the number of molecules, which is directly proportional to the mass of water involved. The amount of energy per unit mass of liquid water is called the latent heat of vaporisation, which is supplied by the kinetic energy of the molecules within the liquid. The actual amount of energy required varies slightly with temperature, which will be discussed in the following section.

Evapotranspiration represents a more complicated use of water molecules by transpiration (the process by which they escape from the leaves to the atmosphere) that causes the plant nutrients in the soil to rise and aid the building of plant tissues. Indeed this process is known as a 'consumptive use' since the water molecules represent a substantial part of the food-fibre production in the plant system. The process also represents the most important movement of water in the hydrological cycle (Figure 6.1) since it accounts for the disposal of nearly 100 per cent of the annual precipitation in arid regions and about 75 per cent

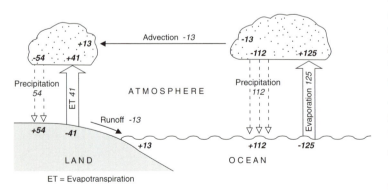

Figure 6.1 *Schematic representation of the average annual hydrological cycle (cm yr⁻¹)*

in humid areas. Transpiration takes place through the leaf openings or stomata, the lenticles or occasional holes in the bark, the surface of spongy leaf cells and the cuticles of leaves. Water molecules are also lost in the soil zone adjacent to the plant since the root system obtains water by absorption at surfaces in contact with soil moisture. The transpiration process depends primarily upon the supply of radiant and sensible heat energy to the plants, which stimulates evaporation within the leaf.

As was mentioned in the introduction to this chapter, the evapotranspiration process takes place as either an actual or a potential water loss. Actual evapotranspiration (AE) is the process by which water vapour escapes from the living plant (principally through the leaves) and enters the atmosphere, plus the water evaporated from the adjacent soil or snow cover. It is controlled by meteorological, soil and plant factors that will be discussed in the next two sections. Conversely, potential evapotranspiration (PE) represents the water need of a plant, which is the supply of soil moisture or precipitation stored on the surface that is at all times sufficient to meet the demands of transpiring plants. PE has been defined as the water loss that will occur if at no time there is a deficiency of water in the soil for the use of vegetation. Since this condition can be maintained permanently only by supplementing natural rainfall with irrigation, PE is almost entirely controlled by meteorological factors, with stomata size and plant type the only other variables concerned.

Meteorological factors affecting evaporation and evapotranspiration

Meteorological factors interact in the boundary layer to provide the energy and diffusion mechanisms that promote water loss (namely evaporation, AE and PE) from all surfaces. As was discussed above, energy is required to break the bonding of molecules in a water body or a thin film covering soil grains and leaves. Consequently, solar radiation (or more correctly net radiation, see Chapter 4) provides this thermal energy, which is known as the latent heat of vaporisation, and is used to effect the change from liquid (or ice) to vapour. The actual amount of energy required decreases as the temperature increases by about 0.1 per cent per degree Celsius since, at higher temperatures, the intermolecular spacing in the liquid is greater and the phase change takes place more easily. For example, the latent heat of vaporisation at 0 °C is 2.501 MJ kg^{-1}, at 20 °C is 2.450 MJ kg^{-1}, and at 30 °C is 2.430 MJ kg^{-1}. Sublimation requires even more energy, since the latent heat of fusion (0.33 MJ kg^{-1}) is needed first to melt the ice crystals before the latent heat of vaporisation is effective. At 0 °C, the latent heat of sublimation equals 2.83 MJ kg^{-1}.

Apart from providing the necessary thermal energy for the initial liquid to gas change, net radiation also adds essential mobility to individual water molecules. This kinetic energy breaks down the intermolecular bonding (discussed above) and increases the separation and velocity of the water molecules. It also increases their chance of passing through the restrictive membrane of the water surface to facilitate their transfer into the adjacent air. As the faster molecules are the first to 'escape', so the average energy and temperature of those composing the remaining liquid will decrease and the amount of energy required for their continued release must become correspondingly greater. As a result, evaporation reduces the temperature of the remaining liquid (by an amount proportional to the latent heat involved) and is a well-tried domestic cooling technique.

It is generally regarded that solar (net) radiation is the most important single factor involved in the flux of vapour from water bodies and plant surfaces. Consequently, all vapour transfers have close relationships with the diurnal and seasonal distribution of solar radiation, being greatest during the day (and especially around solar noon) and in summer (around the solstice month of June in the Northern Hemisphere). Table 6.1 indicates this importance when correlating vapour flux from three evaporimeters (simple pans or tanks full of water) and the main meteorological variables involved. In all cases, solar radiation had a good and significant correlation with the water loss (especially the high 0.77 correlation with the Black Bellani tank).

Temperature has often been regarded as a surrogate for solar radiation when the latter data are unavailable (e.g. in the Thornthwaite estimations of PE in 1948). Since air temperatures are largely dependent upon solar radiation, it is expected that a reasonable correlation would occur between them and the rate of vapour flux. However, lateral and vertical heat transfers in large, deep water bodies (in particular) result in a considerable time lag between maximum temperatures and evaporation rates in diurnal and seasonal terms. Indeed, as discussed later, evaporation rates from very deep lakes are at a maximum in the low-sun winter season, with complete reversal of the expected trend. Consequently, Table 6.1 indicates a weak correlation between temperature and evaporation. It is apparent that temperature differences between the water body and the surrounding air are much more pertinent to evaporative losses than air temperature on its own, since they control the vital vapour pressure deficit. Vapour pressure is the independent partial pressure (i.e. weight) of the gas, which varies both temporally and spatially, depending on the air mass involved (see Chapter 12). The transfer of water molecules from the water body, soil or leaf into the air depends on vapour pressure differences, which allow the molecules to flow from high to low vapour pressure. Thus, for evaporation and transpiration to take place, the vapour pressure at the water or leaf surface must exceed the vapour pressure in the ambient air above. A reversal of this gradient (i.e. with low vapour pressure

Table 6.1 *Correlation coefficients between various evaporimeters and the daily variation of meteorological factors*

Factor	Black Bellani tank	Summerland tank	4-foot tank
Solar radiation	0.771	0.603	0.601
Vapour pressure deficit	0.719	0.695	0.457
Air temperature	0.460	0.421	0.304
Wind speed	0.269	–0.012	0.232

Significance (N = 153 cases): 5% level, r = 0.16; 1% level, r = 0.21
Source: Adapted from Holmes and Robertson, 1958.

existing at the surface) causes vapour flux towards the surface which retards further vapour flux and actually stimulates condensation (see Chapter 7).

A further limiting condition is the absolute vapour saturation limit (referred to as saturation vapour pressure or SVP), since vapour flux cannot continue into a saturated atmosphere. A vapour pressure deficit represents the difference between the SVP at the temperature of the water surface and the actual vapour pressure at the temperature of the surrounding air. The SVP changes dramatically with temperature (e.g. from 1 mb at −22 °C to 100 mb at 46 °C), which controls the number of water molecules that are allowed to 'escape' into the atmosphere. Consequently, when the water body and air temperatures are identical, the vapour pressure deficit is minimal and evaporation proceeds slowly. For example, with a water and air temperature of 30 °C, the rate of evaporation will be 0.41 mm hr^{-1}. Conversely, when the water temperature is 6 °C higher than the adjacent air, vapour pressure deficits are accentuated and the rate of evaporation is doubled to 0.82 mm hr^{-1}. Table 6.1 where the correlation coefficients for two of the evaporimeters were close to the 'good' association of 0.7, confirms the importance of vapour pressure deficit in the evaporation process.

Relative humidity represents the proximity to saturation of the air (see Chapter 7) and refers to the actual moisture level in the ambient air as a ratio of the amount held when saturated at that temperature (*viz.* 100 per cent). Obviously, relative humidity will control the 'thirst' of the air since when it increases (following a temperature decrease), proportionately fewer water molecules leaving the evaporating surface can be held in the air, so the rate of vapour flux is gradually reduced. Conversely, a fall in relative humidity (associated with a rise in temperature) will increase the water-holding capacity of the air, which will increase the rate of vapour flux from the surface. Obviously, daytime warming is conducive to vapour flux, whereas nocturnal cooling brings about the attainment of saturation point and dew point, which favours condensation processes (see Chapter 7). With a water and air temperature of 30 °C, the rate of water loss will be 0.61 mm hr^{-1} with a relative humidity of 40 per cent, compared with 0.41 mm hr^{-1} at 60 per cent, 0.20 mm hr^{-1} at 80 per cent and zero evaporation at 100 per cent.

The accumulation of water vapour in the air overlying a water body or leaf will eventually lead to saturation of the lowest air layers and the consequent termination of evaporation/transpiration. This situation would occur fairly soon in absolutely calm conditions, when air stagnates over the evaporating or transpiring surface and vapour flux continues very slowly (if at all) between air molecules by molecular diffusion. However, in windy and turbulent conditions, vigorous air movements mix the lowest saturated layers with drier overlying air and, indeed, continually advect unsaturated air over the water body to accelerate the vapour flux. This turbulent action is known as eddy diffusion and represents the principle of artificial fanning in hot, humid climates. Perspiration and evaporative cooling, which use up blood heat to cool the body, are accelerated when artificial turbulence/eddy diffusion is induced by the fan, which renews the supply of dry air over the perspiring body.

It is apparent that eddy diffusion only enables vapour flux to proceed at the maximum rate allowed by the other meteorological variables, discussed above, especially solar radiation and vapour pressure deficit. This leads to a poor relationship between wind speed and evaporation in Table 6.1. Indeed, the Summerfield tank correlation coefficient was close to zero, and the negative tendency indicated that evaporation actually decreased with higher wind speeds. It should be noted here that these weak relationships must be partly due to the fact that the very limited fetch of a 1 m tank is completely insufficient to maximise the influence of eddy diffusion, especially when compared with a large lake or ocean. It is clear that wind speed does not appear to influence the actual initiation of the evaporation process but, by removing vapour-laden air, permits a given rate of vapour flux to be maintained. The relationship between wind speed and evapotranspiration appears to be even more tenuous since the crop surface tends to seal itself as wind speed increases, which results in a marked restriction of turbulent mixing with depth. It is apparent that the influence of wind speed on transpiration is not all that important, especially when compared with the influence of solar radiation and vapour pressure deficit.

Environmental factors and the rate of evaporation and evapotranspiration

The previous section emphasises the importance of meteorological factors in initiating and sustaining all the vapour flux processes (Table 6.1), which forces the environmental factors into a more passive role. However, it is obvious that these water, soil and plant characteristics will control the effectiveness of the atmospheric factors involved in vapour flux, although the actual flux rates depend on the process concerned. Table 6.2 summarises these individual contributions and indicates that whereas the rate of evaporation is affected by water and soil characteristics and AE is stimulated by soil and plant features, PE is controlled almost entirely by meteorological factors. In fact, only two plant features (namely stomata size and plant type, which are independent of the permanent water surplus required by this flux) are able to interrupt this atmospheric 'monopoly' of PE. For example, stomata size dominates at night when the stomata are half-closed and net radiation deficits are common.

Water characteristics affect only evaporation and include the quality, depth and size of the water body. The salinity of the water is important since it reduces vapour pressure, and evaporation decreases by about 1 per cent for every 1 per cent increase in salinity. Consequently, evaporation from sea water (with an average salinity of about 3.5 per cent) is 2–3 per cent lower than evaporation from fresh water. Water pollution appears to be an indirect limiting factor, since changes in water colour and turbidity will change the surface albedo. Over time, this will modify the natural energy balance of the water body and change the heat storage, which could alter the normal evaporative losses. As was mentioned earlier, water depth has quite a considerable influence on

Table 6.2 *The contribution of meteorological and environmental factors to the evaporation and evapotranspiration processes*

Process	Atmosphere	Water	Soil	Plant
Evaporation	X	X	X	—
Actual evapotranspiration (AE)	X	—	X	X
Potential evapotranspiration (PE)	X	—	—	X*

*Only stomata size and plant type significant.

evaporation rates, mainly associated with varying heat capacities. Shallow water bodies have a restricted heat storage or thermal capacity, so that the seasonal air and water temperatures are in harmony, which means that maximum water temperatures occur in midsummer, with associated maximum evaporative losses at this time. For example, the Kempton Park Reservoir near London (9 m maximum depth) records about 125 mm evaporation in July, compared with about 12 mm loss in December.

Conversely, deep water bodies have a much higher heat storage, which, coupled with the deep mechanical mixing activated by the winter water 'turnover', releases heat slowly during the low-sun season. As a result, the heat energy released at this time is made available for evaporation at a time when temperatures (and SVP) are much lower in the surrounding air. This develops a vapour pressure deficit, which encourages a large amount of evaporation that can exceed that of the high-sun summer season, when air temperatures (and SVP) are higher. For example, Lake Superior in the USA (397 m maximum depth) records about 120 mm evaporation in December, compared with about 30 mm in June. The size of the water surface also influences the build-up of a protective vapour 'blanket' of saturated air, which decreases the rate of evaporation as it thickens. Large surface areas have the greatest 'blanket' development, which will reduce the depth of water evaporated.

Soil factors control the effectiveness of evaporation and actual evapotranspiration since they determine the concentration of water molecules in films surrounding the soil grains and filling the spaces between them. These molecules escape into the atmosphere, when motivated by the meteorological factors described earlier, by direct vapour flux or indirectly through transpiration via the root–stem–leaf system of the plant. The main controlling factor must be the soil moisture content in the surface layer, since this represents the potential for water availability or the so-called evaporation opportunity of the soil material. However, the relationship between soil moisture content and evaporation/transpiration is quite close, proceeding at a 100 per cent 'evaporation opportunity' when the soil is saturated (at field capacity) and decreasing rapidly as the surface moisture content falls until it is zero with a dry soil surface.

It is apparent that the moisture content of the top few centimetres of surface soil is the decisive factor, since subsoil water may not affect the rate of surface vapour flux from shallow-rooted plants. In dry areas, however, the role of subsurface water is influenced by the soil capillary features, which are responsible for the upward movement of soil water (against gravity) through the capillary tubes from wetted to dry grains (i.e. towards the dry surface soil). This movement is greatest (over a metre or so) in fine-grained soils, compared with only a few centimetres rise in coarse-textured material. Of course, the significance of subsurface water is increased when the water table (or upper level of subsurface water) is close to the surface, increasing the potential of groundwater availability for the 'evaporation opportunity'. Surface evaporation reaches maximum amounts when the water table is at the surface, averaging 5 mm day^{-1} at that time, compared with 0.5 mm day^{-1} when the table is at a depth of 140 cm. It appears that evaporation decreases quite rapidly initially as the water table retreats downwards. However, after a depth of about 1 m has been reached, any further decrease in water table height is accompanied by only a slight change in the evaporation rate.

Additional soil factors include the colour of the material, which determines the surface albedo and heat storage in the soil, which represents the potential for the latent heat of vaporisation. For example, dark soils with an albedo below 10 per cent should absorb maximum amounts of solar radiation and record highest surface temperatures and maximum amounts of evaporation. The degree of exposure of the soil surface is also important since, without a protective cover of vegetation, evaporation will proceed at an accelerated rate. For example, evaporation from bare soil is about twice as fast as that from soil under forest cover.

The last set of environmental controls involves plant factors, which naturally are only associated with transpiration rates. Transpiration is vital as a life function in that it causes a rise of plant nutrients from the soil and cools the leaves. The concentration of solvent molecules in cells of the plant roots exerts an osmotic tension of up to 15 atmospheres upon the water films between the adjacent soil grains. Furthermore, as these water films shrink, the tension within them increases and, if the tension exceeds the osmotic root tension, then the continuity of the plant's water supply is broken and wilting occurs. Four plant factors contribute to the transpiration losses from vegetation, *viz.* stomata size, plant type, plant colour and stage of growth. All four determine the effectiveness of AE, but PE involves only the first two. Stomata are small pores in the leaves that are open during daylight. At this time, the actual pore size does not affect the rate of transpiration and, indeed, the vapour flux is governed by the prevailing meteorological factors. However, at night, the stomata are half-closed, and pore size becomes the dominant factor controlling the nocturnal vapour flux.

The type of plant determines the availability of moisture and the degree of transpiration motivated by the atmospheric factors. Vascular and non-vascular plants are extreme cases of water potential, which is severely limited in the latter case. Non-vascular tundra plants, for example, are resistant to vapour flux and contribute largely to the polar desert environment. The colour of the plants controls the albedo of the vegetation and associated heat storage in the plant fabric, which is responsible for the latent heat available for transpiration. Light-coloured plants with a relatively high albedo (above 20 per cent) reflect a considerable amount of solar radiation, and consequently have lower leaf temperatures, with minimum amounts of transpiration. The stage of growth of a plant will also determine the potential for moisture supply and the associated availability of water for transpiration. For example, maize is green in the early growing season, containing maximum amounts of plant water, which encourages excessive amounts of transpiration under favourable meteorological conditions. Conversely, at the harvest stage, the crop is brown and desiccated, with a greater resistance to vapour flux.

CASE STUDY 6: Suppression of vapour flux by deliberate human activities

About 26 per cent of the terrestrial surface of the Earth is water-deficient, and modifications of the hydrological cycle (Figure 6.1) are needed to meet the water demands of these arid lands. This long-term recognition is now coupled with the problem of increasing desertification that is often associated with short-term climate change and more frequent and severe droughts. Furthermore, this problem has been aggravated by overgrazing and non-traditional agricultural practices (such as cash crop farming), which were introduced into semi-arid lands (like the African Sahel) when rainfall regimes were adequate to support local food production. Acute population increases in these regions (e.g. 4 per cent per annum in Kenya) are placing excessive demands on medium- to low-potential land in semi-arid areas in order to maintain adequate levels of food production. However, since water deficiency is recognised as a key limiting factor for stabilised and improved agriculture in these regions, the case for deliberately modifying the hydrological cycle becomes paramount.

The most obvious and common modification of this cycle concerns the condensation process (see Chapter 8), since most of problems of the semi-arid lands occur when rain fails. The case study at the end of Chapter 8 examines the techniques and success of these modifications, mostly associated with cloud-seeding practices. Alternative techniques are discussed here that can be introduced to maximise the efficiency of existing water levels in lakes and plants by reducing natural rates of evapotranspiration. The traditional planting of cross-wind barriers (Figure 6.2) is an accepted way of modifying airflow patterns, although near solid barriers do introduce strong eddying. More open shelterbelts (Figure 6.3) can, however, prevent these eddies but can still reduce the structural damage and wind chill from a strong airflow (e.g. in the Rhône Valley, southern France, as protection from the notorious Mistral wind, which is discussed in Case Study 12). Measurements have revealed (Figure 6.3) that open windbreaks can reduce air velocity in the lee of the windbreak by 70 per cent at a distance away of five times the height of the break. The associated eddy diffusion (discussed earlier as a factor influencing the effectiveness and rate of evapotranspiration) is also reduced by 15 per cent at this distance, increasing the efficiency of soil and plant water. However, windbreak effectiveness decreases rapidly away from the break location (Figure 6.3) to a mere 20 per cent reduction in air velocity at a distance 15 times the height of the wind break. Consequently, regular rows of trees (especially the ideal Lombardy poplars), which characterise the lower Rhône Valley, are necessary to maintain this effectiveness, although this must be at the expense of soil moisture and farming mechanisation.

Alternative schemes have been designed to suppress evaporation and evapotranspiration rates in an attempt to reduce the vapour flux from water

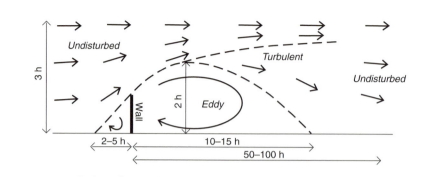

— — Surface of separation

Figure 6.2 *Characteristics of the airflow pattern due to a near-solid cross-wind barrier (not to scale) (after Gloyne, 1954).*

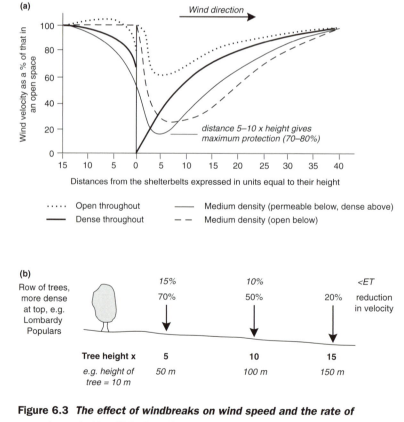

Figure 6.3 *The effect of windbreaks on wind speed and the rate of evapotranspiration ((b) after Gloyne, 1954).*

bodies. Natural vegetation cover (e.g. duckweed and water lilies) has been known to suppress evaporation rates by 20 per cent. Consequently, an artificial covering of non-toxic chemical films (e.g. cetyl alcohol or hexadecanol) can be used with similar effect. Under calm conditions, these films have reduced evaporation by as much as 9 per cent from large lakes and 25 per cent from small ponds, at a cost of US$40 per 1000 m^3 of water saved (at 1976 prices). However, their use is quite ineffective in the overall agricultural calendar, since the films ultimately break down and are rapidly dispersed by winds. Evaporation rates can also be reduced with the 'correct' design of reservoirs with regard to their depth and size, as was discussed earlier. Deep reservoirs and lakes (e.g. Lake Superior, USA) have lower surface temperatures and winter peaks of evaporation (i.e. in the low-demand season) and large water bodies have an increased vapour blanket and lower vapour flux rates.

Evapotranspiration rates can be suppressed when plants are sprayed with non-toxic chemicals that partially close the stomata. For example, atrozine (applied to crops at a rate of 20 ppm) has effectively reduced the vapour flux from corn by 45 per cent and soya beans by 70 per cent within six hours of application. This scheme is operational in the conifer plantations at Yattendon, near Reading, England, when applications are made every October to reduce transpiration rates and the needle drop from Christmas trees. However, the suppression of transpiration seriously limits plant growth by reducing the supply of soil nutrients that aid the building of plant tissues, which represents a substantial part of the food-fibre production in plants.

7 The condensation process and condensation forms at or close to the Earth's surface: dew, hoar frost and fog

Condensation can be regarded as the second pathway of the hydrological cycle, since it is responsible for the return of water vapour to a liquid or solid state. This chapter covers:

- **definitions of terms used**
- **explanation of air saturation**
- **the attainment of saturation by cooling mechanisms**
- **discussion of the formation of dew, hoar frost, radiation fog and advection fog**
- **case study: fog dispersal and frost protection by human activities**

Box 7.1

Condensation processes, terminology and their role in global water exchanges

1   The term 'condensation' strictly applies only to the process by which water vapour is changed into liquid water.

2   The change from vapour to a frozen state (omitting the liquid stage) is more correctly termed sublimation, although the term is also used for the change from a frozen state to vapour (see Box 6.1).

3   Together with evaporation and evapotranspiration, condensation is responsible for colossal amounts of water exchange over time, in the order of 19 million tonnes a second.

4   Even so, if all this atmospheric water were suddenly to be released as precipitation it would amount to only about 25 mm deposited over the entire Earth's surface.

5   In other words, the atmosphere contains the equivalent of only 25 mm of precipitation at any one time. Since the average annual global precipitation equals 915 mm, then there must be 36 evaporation/condensation cycles each year.

6   This means that a molecule of water vapour will remain in the atmosphere for an average of 10 days, drifting up to hundreds of kilometres before it condenses and precipitates back to the Earth's surface.

The condensation process

As long as the vapour pressure of the air is less than the SVP (see Chapter 6) over a liquid or frozen surface, then a vapour deficit exists, which favours a vapour flux. However, when the current vapour pressure equals the SVP, then a state of equilibrium will occur. This prohibits evaporation and condensation, since the rate of molecular transport across the water–air interface is the same in both directions. Finally, when the existing vapour pressure of the air exceeds the SVP, then the vapour flux is towards the surface, which leads to condensation.

Table 7.1 *Saturation vapour content and air temperature*

Temperature (°C)	SVC (g m⁻³)
−1.1	4.4
4.4	6.5
21.1	18.3
26.6	25.0

When unsaturated air rests on free water, soil or plant surfaces, the water steadily evaporates until the concentration of water vapour in the atmosphere (i.e. the humidity of the air) reaches saturation level. This level is maintained by the fact that there is a strict limit to the amount of water vapour that can be held in the air. After so much water has been evaporated into the confined space of the atmosphere, any further evaporation will produce an equal amount of condensation, so that the amount of vapour present in the air will remain constant. This maximum value is referred to as saturation and the saturation vapour content (SVC or water vapour capacity) of the air is controlled primarily by the temperature of the air (Table 7.1). Cold air has a very limited capacity and, even at saturation point, it contains very little moisture. However, as the temperature rises, the capacity of the air for water vapour increases rapidly and, furthermore, the rate of capacity increase itself grows greater as the temperature rises. In other words, a temperature rise of 5 °C is more than three times as effective at 21.1 °C as at −1.1 °C in increasing the amount of water that can be held in the air.

It is apparent that saturation is achieved by increasing the amount of water vapour in the air by continued evaporation/transpiration until the current vapour content equals the SVC at that temperature and the saturation deficit is eliminated. Alternatively, if the temperature falls then the SVC will decrease accordingly until it is just equal to the actual amount of water vapour held by the air, thereby achieving saturation. This is the most important natural process which leads to the saturation of air. For example, at 26.6 °C and a current vapour content of 18.3 g m⁻³ (i.e. a 6.7 g saturation deficit), saturation will take place when continued vapour flux increases the vapour content to the SVC (25.0 g m⁻³) or the temperature can be lowered to 21.1 °C.

If the temperature continues to fall below the point at which the air becomes saturated, then there will be an excess amount of water vapour present compared with the SVC of the air at the new, lower temperature. Consequently, this excess vapour will be released by changing its state through condensation to a liquid or solid form, and the temperature that forces this change is called the dew point (see Chapter 5). In Table 7.1 for example, when the temperature falls from 26.6 to −1.1 °C, then

20.6 g m^{-3} of water vapour must be released through condensation to maintain the 4.4 g m^{-3} SVC at the lower temperature (*cf.* 25 g m^{-3} at 26.6 °C). Since the cooling of air is vital for the attainment of saturation, it is necessary to consider the reasons for this temperature decrease. The three major processes that combine to cool air to and below the dew point are as follows:

1 Contact cooling when air is chilled by conduction from a surface cooled by long-wave terrestrial radiation.
2 Air mass mixing, especially when warm, moist air is chilled by conduction during advection across a cold surface.
3 Adiabatic cooling as air parcels rise and expand in unstable situations and the temperature falls by the DALR and eventually the SALR

However, before the cooling of saturated air can condense vapour into a variety of weather phenomena, non-aqueous hygroscopic nuclei (i.e. atmospheric particles that have an affinity for water) must exist in the air. These nuclei are essential for condensation since, in their absence, the air can become supersaturated and the relative humidity can exceed 100 per cent. Laboratory experiments have indicated that pure, saturated air can be cooled to approximately –40 °C without condensation/sublimation occurring. Below this temperature, water vapour appears to sublime spontaneously to ice crystals in the absence of transient impurities. Conversely, a large concentration of nuclei can promote condensation at 70 or 80 per cent relative humidity, which partly accounts for the smog characteristics of urban and industrial areas.

Hygroscopic nuclei vary widely in composition, although salt (sodium chloride) and combustion products (sulphur trioxide) have the best water-seeking properties. Salt occurs in the form of large nuclei with radii greater than 1.0 μm and is very widespread, since it enters the atmosphere following the evaporation of ocean spray. However, it is present in small concentrations, usually less than 10 cm^{-3}. Combustion products are characterised by small nuclei (with radii less than 0.1 μm), which accumulate in the air following volcanic eruptions and the burning of sulphur-containing fuel like coal and oil. Consequently, their main distribution is restricted to urban and industrial air masses, where large volumes of sulphur products are released. However, despite their limited spatial distribution, they are present in large concentrations of up to 10^{6}cm^{-3}. The enormous number of dust particles entering the atmosphere from arid lands is believed to be of secondary importance as condensation nuclei, since the majority of the particles are non-hygroscopic, with a very weak water affinity. The size and concentration of all nuclei in the atmosphere determine the dimensions of water droplets forming on them, which influence fog and precipitation characteristics, as described in the next section and Chapter 8.

Condensation forms at or close to the Earth's surface

Saturated air with an abundance of hygroscopic nuclei will condense at or close to the Earth's surface when the air is chilled below its dew point by the contact cooling and air mass mixing described earlier. The former cooling mechanism is responsible for the formation of dew, hoar frost and radiation fog and is associated with decreasing ground temperatures following the loss of heat by long-wave terrestrial radiation (see Chapter 3). This infrared heat loss to the atmosphere is accelerated on calm, clear nights and is confined to a very shallow layer of air in direct contact with the cold ground, since air is a very poor conductor and the cooling process is essentially by conduction. On cooling below the dew point, the excess water will condense in the air as radiation fog or be deposited on the surface of objects as dew, or as hoar frost when the object's temperature is below freezing point with sublimation directly from the water vapour. The formation of dew and hoar frost is most uneven, since dark objects (like vegetation, thermally insulated from soil heat storage by its foliage) always cool rapidly. They behave as perfect radiation black bodies (i.e. good absorbers and radiators), which, at night, provide the coldest surfaces and encourage the greatest deposition of droplets or ice crystals.

The conditions that lead to the formation of dew and hoar frost are clear, calm nights with moist air close to the Earth's surface (i.e. with a relative humidity preferably above 80 per cent at sunset). Clear, cloud-free conditions are necessary to allow maximum long-wave radiation loss and ground cooling, since a 'blanket' of cloud will facilitate a greater 'greenhouse effect' and surface warming by counter-radiation (see Chapter 3). Windy nights prevent the warm, moist air from stagnating and remaining in contact with the chilled Earth or cold objects long enough to cool sufficiently, even with clear skies.

Radiation fog forms under the same conditions required by dew and hoar frost when, on clear nights, the Earth's surface cools rapidly by long-wave radiation loss and chills the surrounding moist air by conduction. However, the air has to be cooled below its dew point over a longer period of time and a much greater depth than the 'skin' of air surrounding cold objects that is necessary for dew formation. Under these more advanced conditions, condensation will now form as dew on surface objects and on the hygroscopic nuclei of the adjacent air (normally up to 300 m). This condensation yields minute droplets suspended in the lowest 100–300 m of atmosphere, and since the cooling of the air depends on ground chilling below by long-wave radiation, the resultant droplet accumulation is termed radiation fog. Whereas calm, stagnant air is necessary for dew formation, radiation fog requires a light wind (up to 3 m s^{-1}) to stir the cold air in contact with ground gently and to give sufficient turbulence to spread the cooling upwards over a greater depth of air. Consequently, this turbulence thickens the fog from below when dew-point chilling and droplets are diffused aloft. However, above 3 m s^{-1}, the vigorous airflow would prohibit the necessary air stagnation and contact cooling

Plate 7.1 *Radiation fog near Llangollen, North Wales, UK.*

beyond the critical dew-point level. Fog can exist at temperatures below 0 °C when the droplets are supercooled and freeze on to cold objects to form fragile, low-density rime ice.

Radiation fog forms first and thickest in valleys or hollows (like the Thames Valley in southern England) since cold, dense air drains downslope by gravity and collects at the lowest elevations (Plate 7.1). This katabatic airflow is described in detail in Chapter 16 and is responsible for the characteristic valley fog that forms during stable anticyclonic weather, when the tops of telegraph poles and church steeples sometimes rise out of the cold, raw fog blanket. Radiation fogs are also thickened by temperature inversions (see Chapter 5), when cold, dense air is trapped beneath a 'lid' of overlying warm (and fog-free) air. The fog's top now lies near the base of this inversion at a height of several hundred metres, with the crests of hills projecting upwards through the fog into clear, warm, dry air above. Consequently, radiation fog is normally extremely patchy, preferring sheltered valley locations, although it occasionally can become more widespread and regional.

After sunrise, the radiation fog is normally 'burned off' quite rapidly by terrestrial heating and evaporation from the warming surface. The fog evaporates from below and gradually lifts during the early morning until the last upper remnant has the appearance of low, patchy stratus cloud extending outwards from the high ground. However, the evaporation and elimination of radiation fog after sunrise can be delayed in thick, high fogs forming under strong temperature inversions. Furthermore, when the concentration of smoke and aerosols (released by urban and industrial sources) destroys the transparency of these fogs, then the resultant opaque fogs can persist for many days as serious smog episodes. Perhaps the best example of this was the infamous smog over London, England, which developed in the stable weather of a winter anticyclone in early December 1952. The fog build-up was accentuated by the accompanying strong temperature inversion and fumigation-type pollution (see Chapter 5), and the resultant smog persisted for five days. The concentration of smoke and sulphur dioxide reached a peak on 7 December, at 1.6 mg m^{-3} and 0.75 ppm respectively, and the number of deaths reached a maximum (about 900) on the following day, totalling 4000 over the period.

Condensation associated with air mass mixing produces advection fog, which develops when cold or warm airstreams move across warm or cold surfaces, respectively (see Chapter 12). The first type of advection fog is related to the passage of cold polar air across a warmer sea surface, which mixes with the warm, moist air, providing a cooling–condensing mechanism. The warm vapours evaporating from the water body, concentrated in a shallow 'skin' or 'envelope' at the surface, are immediately chilled by conduction with the cold airstream and condense into a very shallow zone of fog. In fact, the sea surface appears to steam or smoke, giving rise to names like steam fog or Arctic smoke. It is very common in polar areas, where very cold air drains off a snow-covered land mass over the warm water of exposed leads or pools in the sea ice.

The second type of advection fog is associated with the passage of warm, humid airstreams over a colder surface, especially a cold ocean current (see Figure 12.1b). It is the most common type of marine fog encountered, since about 80 per cent of all such fogs owe their origin to this process, and its formation depends on a significant contrast between air and water temperature, and a wind speed (and related turbulence) of about 4 m s^{-1}. On such occasions, the warm, humid air is transported across the cold water and is chilled from below by conduction, leading to the development of a strong temperature inversion and stability (see Chapter 5). Conductive cooling below the dew point causes condensation in the surface air, and the necessary turbulence spreads the cooling upwards over a greater depth of air. However, if the airflow is too vigorous, exceeding 8 m s^{-1}, it will lift the fog to form a low stratus cloud with a base of 100 m or so.

These extensive marine fogs are particularly well-developed in the Grand Banks sea area to the east of Newfoundland, Canada, where warm, humid south-easterly winds from over the Gulf Stream encounter the southward-flowing cold Labrador Current. The thermal contrasts and associated fog development are most pronounced in the summer months, when they dominate on two out of three days. These marine fogs also dominate California (Plate 7.2) in the San Francisco area, when warm, moist south-westerly winds flow onshore across the cold California Current, which represents the upwelling of cold benthos water close inshore. Advection fogs can also be found inland when moist, warm air flows over a cold land surface, especially when the surface is covered by thawing snow close to 0 °C. These fogs are more extensive than radiation fogs and will persist over large areas as long as the airflow is maintained. Over northern land masses, the more extreme winter cold (−30 °C) causes advection fogs to be composed entirely of ice crystals. Here, sublimation forms ice fog or 'diamond dust' with accompanying rime-ice deposition and surface optical phenomena (i.e. mock suns and sun pillars) due to sunlight refraction through the crystals.

Plate 7.2 *Advection fog near Santa Barbara, California, USA, taken by M. Turnbull.*

CASE STUDY 7: Fog dispersal and frost protection by human activities

Even though modification of the pathways in the hydrological cycle has received most attention, schemes to disperse fogs at airports and to protect crops from frost have been much more successful and operational. Indeed, both practices can result in economic advantages and increased profits for the commercial success of airport authorities/airlines and farmers. Fog dispersal operates at leading airports around the world, and the actual technique employed depends entirely on the thermal structure of the fog concerned. Indeed, two types of fog are prevalent, namely the cold (supercooled droplet/ice crystal) fogs of high-latitude/continental airports in winter (e.g. Montreal, Canada; and Moscow, Russia) and the warm (water droplet) fogs of low-latitude/maritime locations (e.g. London, England; Paris, France; and San Francisco, USA). The formation and characteristics of these fogs were discussed above, when radiation chilling and advective mixing brought about the attainment of dew point and condensation (droplet fog) or sublimation (ice fog or diamond dust). However, it should be noted that radiation fogs are more localised and short-term (nocturnal) in nature and, indeed, can be avoided by careful airport siting (i.e. the favoured katabatic control (see Figure 16.1b) will not be as great a problem outside a valley location and inversion hollows, where major airports are obviously not sited). Conversely, advection fogs are much more extensive regional developments and will persist for days over large flat areas (i.e. ideal airport sites) as long as the advected airflow is maintained. The dispersal of warm fog is a much easier proposition, since it is a localised and temporary phenomenon and can be removed by the addition of heat/turbulence to accentuate the evaporative process (see Chapter 6). For example, an operational scheme at a Paris airport (France), appropriately named turboclair, uses old jet aircraft engines sited along the main runway.

During fog episodes, the engines are fired to produce 700 °C heat, which is sufficient to clear the fog in the vicinity by evaporation processes. Problems associated with this significant source of heating include the emission of pollutants in the exhaust gases and increased air turbulence, which interferes with the landing of light aircraft (like the Cessna 310). Cold fogs cannot be removed by attempts to accelerate evaporation and indeed have to be tackled by dry-ice seeding techniques, which are fully explained in the case study at the end of Chapter 8. Dry ice (the common name for carbon dioxide crystals) acts as freezing nuclei for the supercooled droplets suspended over the runways. Helicopters can hover over these shallow fogs and distribute the dry ice into the cold fog. This artificially stimulates the Bergeron–Findeisen process (see Chapter 8 and Figure 8.2) and the fog simply precipitates on to the runway, clearing the air and allowing normal airport operations. Precipitated ice crystals can quickly be removed by snow ploughs, which are readily available at these high-latitude/ continental locations.

Farmers and horticulturalists are plagued by unseasonal frosts, particularly in springtime, when delicate young plants and fruit blossom are most vulnerable. Severe radiation frosts at this time can decimate the crops concerned and will induce financial ruin and annoying shortages (e.g. in the Kent, England, apple orchards in April 1997). A number of schemes are available to protect against unseasonal frosts, although they are mainly small-scale and limited to a particular field or orchard. Most of these schemes are designed to counteract the basic requirements for radiation frosts, namely strong surface cooling on clear, calm nights (see above). The use of water sprinklers is a cheap and easy way to protect against ground frosts and light air frosts. First, the soil is saturated, which increases its thermal conductivity and extends the cold wave to a great depth (see G flux in Chapter 4 and Figure 4.1) thus ameliorating the frost's impact. Furthermore, when this water freezes it releases the latent heat of fusion (0.33 MJ kg^{-1}), which again moderates the chilling.

Suspended canopies, mulching and smokescreens can all prohibit the long-wave radiation heat loss from the ground at night and indeed can act as a very effective, artificial greenhouse effect (see Case Study 3). Surface cooling is moderated and air frost damage can be avoided by these simple, inexpensive schemes, although the scale involved is small and smoke damage to crops could be a problem. The traditional way of combating air frost, on a field or orchard scale, is to produce heat from oil burners (termed smudge pots), which both warm the air and promote air turbulence (and destroy the required air stagnation, see above). Experiments in a 6 hectare citrus orchard burned 150 litres of oil, using 112 heaters per hectare, and raised temperatures by 5 °C within 10 m of the Earth's surface. The effectiveness of this modest heat source was due to the presence of a strong surface temperature inversion, which normally accompanies radiational cooling (see Chapter 5 and above). The inversion prevents the upward movement of warmed air and confines it to the boundary layer, where crops and fruit trees are located, although the associated smoke pollution could be a problem.

The existence of warm air aloft in a temperature inversion (see Figure 5.1b) is again a source of heating for a boundary layer chilled by radiation cooling, if this heat could be brought down to the ground and mixed with the cold surface air. Large wind machines or aerogenerators are used to create this necessary vertical turbulence, and on nights when an adequate inversion exists, the mixing of warm air aloft with cold ground air was observed to raise temperatures within the growing area by 2 °C after only 35 minutes of machine operation. It must be reiterated here that the success of smudge pots and aerogenerators is confined to radiational frosts with adequate temperature inversions. Frosts caused by the invasion or advection of freezing air cannot be moderated in these ways, since there is no warm air aloft (i.e. no surface temperature inversion). Consequently, heat produced at the surface can escape away from the growing area into the troposphere, and aerogenerators now merely mix cold air at the surface with colder air aloft. Fortunately, advection 'freezes' are rare occurrences in major agricultural areas during the growing season, but they have been known to decimate the citrus orchards of northern Florida during an abnormal southerly outburst of Alaskan air (see Case Study 12).

8 ▶ Condensation forms in the troposphere, away from the Earth's surface: clouds and precipitation

Condensation within the troposphere is responsible for interesting and dramatic weather phenomena such as clouds and precipitation. This chapter covers:

- the attainment of dew point in rising air parcels
- the development of clouds and a classification of major cloud types
- the growth of hydrometeors and the role of terminal velocity in the precipitation process
- precipitation forms in unstable and stable air
- case study: cloud seeding and precipitation modification

Box 8.1

Tropospheric condensation forms and processes

1 The forms of tropospheric condensation are represented by clouds and precipitation and are associated with the adiabatic cooling of rising air parcels in unstable conditions.

2 This cooling process was discussed in detail in Chapter 5 (Figures 5.2b and c) and occurs when parcels of air are forced upwards by free or mechanical convection and expand/cool independently of the surrounding atmospheric environment.

3 Condensation takes place when the dew point is reached during the air parcel ascent, assuming sufficient hygroscopic nuclei (salt or sulphur trioxide; see Box 1).

4 At the dew point, the DALR gives way to the SALR, where the cooling by expansion in the air parcel is opposed by the latent heat released by condensation (see Chapter 5).

5 The dew point elevation is also known as the lifting condensation level (LCL) and represents the base of the cloud (see Figure 5.2).

6 The air parcels will continue to rise and accentuate cloud development as long as instability exists (i.e. the ELR exceeds the SALR, as discussed in Chapter 5).

7 The top of the cloud will indicate an eventual change back to stability when the air parcel is cooling at a faster rate than the surrounding environmental air (see Figure 5.2b).

Cloud forms and development

Clouds result from the condensation of buoyant air parcels, when rising air currents keep the droplets and ice crystals in suspension, since they are very small and light. For example, the average size of a cloud droplet is 0.05 mm, and this growth takes

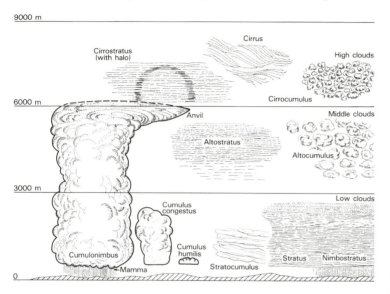

Figure 8.1 *Principal cloud forms and main sub-types*

place in about 100 seconds following condensation. The distinctive base of the cloud is maintained at a uniform level because, if droplets fall below this dew-point level where the air is not saturated, then these small droplets are readily evaporated. When condensation occurs at sufficiently high altitudes (normally above 6000 m), where temperatures are below 0 °C, then the vapour will sublimate directly into ice crystals. In the most basic classification, three principal cloud forms are recognised (*viz.* cirrus, cumulus and stratus) with a number of main sub-types (Figure 8.1). Cirrus are high, thin ice-crystal clouds in delicate feathery filaments or fibrous sheets (cirrostratus), although small ripples occur with turbulence and are called cirrocumulus or a 'mackerel' sky (Plate 8.1). All cirrus/cirro-clouds are associated with the mass advection of ice crystals above 6000 m, and refract the sun's rays to form haloes, mock suns, sun pillars, etc.

Plate 8.1 *Cirrocumulus cloud over eastern Baffin Island, Northwest Territories, Canada.*

Cumulus are flat-based individual cloud masses with a pronounced vertical growth and rounded tops and, when they are associated with instability, they become 'heaped' congestus types (see Plate 5.3). Eventually, cumulonimbus clouds form when the highest part are composed of ice crystals, with the distinctive 'anvil' formation due to strong upper winds, and the cloud base is deformed by strong perturbations into mamma-type undulations (Plate 8.2). Non-buoyant parcels have weak vertical growth and are called cumulus humilis (Plate 8.3) or 'fair weather cumulus'. Compressed cumulus ripples between 3000 m and 5000 m are called altocumulus (Plate 8.4), whereas at lower

Plate 8.2 *Mamma cloud at the base of a cumulonimbus cloud over Guelph, Ontario, Canada.*

Plate 8.3 *Cumulus humilis clouds over Alsace, France, taken by M. Parry.*

Plate 8.4 *Altocumulus cloud over the Antarctic Peninsula.*

elevations (below 2000 m), the larger compressed globular masses are called stratocumulus. Stratus are extended sheets or shallow layers of droplets (or ice crystals when cirrostratus) covering most of the sky. They are associated with the mass advection of droplets under stable conditions below 2000 m (stratus) and between 3000 m and 5000 m, where the cloud is called altostratus. Eventually, a thick mass of stratus/altostratus/cirrostratus (over 10,000 m extent) will produce precipitation to become nimbostratus.

Only the basic cloud types have been described in this section and for a complete, authoritative study of all species and varieties, reference should be made to the *International Cloud Atlas* (WMO, 1956). However, the actual cloud form depends on the free or forced convective mechanism in operation (see Chapter 5) and whether stability or instability exists. For example, with free convection operating under stable conditions, vertical motion is prohibited either before the dew point is reached or soon after, leading to flattened cumulus humilis (Plate 8.3). Conversely, with instability up to the tropopause or a high-level inversion at about 6000 m, free convection proceeds to develop clouds of great vertical extent such as cumulus congestus (Plate 5.3) and eventually cumulonimbus. Forced convection is mainly associated with vertical motion related to orographic or frontal displacement and the mass convergence of air in low-pressure systems, where the actual cloud forms are mainly influenced by the strength of the uplift. For example, when air is forced to rise rapidly over a steep mountain slope, the strong vertical motion causes cumulus clouds to form that will proceed to congestus or cumulonimbus types if instability exists. Conversely, uplift over a gently sloping mountain flank produces a more gradual ascent, which favours the formulation of layered stratus clouds, and altostratus above 2000 m elevation.

The development of clouds along a front depends mainly on the buoyancy of the warm air that is being displaced (see Case Study 14 and Figure 14.6). For example, at a warm front with non-buoyant air, a considerable thickness of layer clouds dominates, including cirrostratus, altostratus, stratus and (eventually) nimbostratus. This development is analogous to the gradual uplift over gently sloping mountain flanks described above. On the other hand, if the displaced warm air becomes increasingly buoyant, then the ascent could lead to conditional instability, which encourages the development of cumuliform clouds, possibly cumulonimbus. Similarly, cold frontal cloud development is controlled by the equilibrium tendency of the rising warm air. With instability, cumulus clouds form rapidly and develop into towering congestus or cumulonimbus (which compare favourably with the cloud formation over steep mountain slopes). Conversely, if the displaced warm air is associated with absolute stability, then cloud development is restricted to less than 3000 m and stratocumulus clouds are more common (see Figure 14.6).

Precipitation forms and mechanisms

Cloud development in the troposphere will ultimately lead to precipitation since the essential difference between cloud and rain droplets is one of size and weight, in terms of the actual strength of the vertical motion. Whether or not droplets, ice crystals and hailstones will be precipitated as hydrometers or will remain suspended in the air as cloud, depends entirely upon their weight in relation to the velocity of the upward currents of air which support them. For every size of droplet or ice crystal, there is a velocity of air which is just capable of holding it at its own level. This is known as its terminal velocity, which varies considerably as the radius/size of the hydrometeor changes (Table 8.1). Since clouds are formed in upward-moving air currents, sometimes of considerable velocity, it follows that there must be processes within clouds which lead to the growth of droplets and ice crystals to a size large enough to overcome the lifting forces and fall to the Earth's surface by gravity.

Table 8.1 indicates that the size of the largest raindrop (2.50 mm) is about 50

Table 8.1 *Terminal velocity of hydrometeors*

Hydrometeor	Radius (mm)	Terminal velocity (m s⁻¹)
Cloud droplets	0.05	0.25
Drizzle	0.25	2.0
	0.50	3.9
Raindrops	1.50	8.1
	2.50	9.1
Snowflakes	5.00	1.7
Graupel	5.00	2.5

Source: Adapted from Pedgley, 1962.

times the size of the average cloud droplet, and it is apparent that special processes must operate in a cloud from which hydrometeors are released. These four growth mechanisms are as follows:

First, coalescence of droplets within clouds when large droplets (0.09 mm), usually forming around large hygroscopic nuclei, fall with greater velocity than small droplets around the smallest nuclei (0.001 mm), since their terminal velocity is considerably greater (i.e. 70 cm s^{-1} compared with 0.01 cm s^{-1}). The large droplet descending through a cloud of smaller ones 'sweeps up' and coalesces with a large number of them lying in its path. The fusion takes place at the rear of the falling large droplet, where the vortex is greatest, and the final size to which a drop will grow in this way depends on a number of basic conditions that determine the actual rate of growth. One of the main factors concerns the fusion efficiency of droplets, which is influenced by their size (i.e. the minimum radius required is 0.02 mm, which is more common in maritime clouds). Also, the liquid water content of the cloud should be large (at least 1 g m^{-3}) and the updraught of air must be moderate (1–5 m s^{-1}), not exceeding the terminal velocity of the droplet. Finally, the cumulus cloud concerned should have an adequate thickness of several thousand metres and should persist for at least 30 minutes.

Second, aggregation of ice crystals (with branching plates) forms snowflakes when they interlock on collision, because of their complex stellar and dendritic branching structures. It appears that this process is aided by the freezing and cohesion of supercooled water-like films on the surfaces of the crystals, which join together when the surfaces come into contact. This cohesion diminishes as both temperature and the degree of supersaturation decrease, so that aggregation occurs most readily at temperatures between 0 °C and –4 °C. Furthermore, the process is very common in clouds composed largely of ice crystals, such as cirrostratus and the uppermost parts of cumulonimbus.

Third, the Bergeron–Findeisen process which is associated with the coexistence of both supercooled droplets and ice crystals for hydrometeor growth, operates in clouds with temperatures below 0 °C. In simple terms, the role of ice particles is associated with their greater attraction to wandering water vapour molecules (due to the resultant supersaturation and deposition on to the particles), compared with that exerted by supercooled water droplets. However, the actual ice crystal growth is accomplished by simultaneous evaporation from supercooled droplets and sublimation–glaciation upon the ice particles as a result of related saturation vapour pressure differences. Figure 8.2 illustrates this growth mechanism, where the vapour flux from droplet to ice particle is stimulated by the gradient associated with the low saturated vapour pressure vortex over ice and high-pressure divergence over the droplet. For example, at –12 °C the corresponding saturation vapour pressures are 2.128 mb over ice and 2.398 mb over the droplet, and this represents a 'peak' flux gradient.

This theory is supported by the fact that showers become most severe when cumulus congestus clouds grow into cumulonimbus with their characteristic ice crystal tops

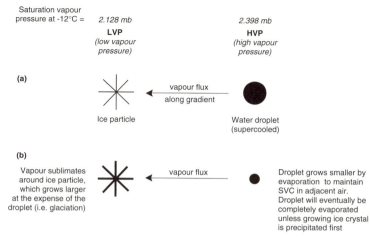

Saturation vapour pressure at -12°C = *2.128 mb*
LVP
(low vapour pressure)

2.398 mb
HVP
(high vapour pressure)

(a)

vapour flux along gradient

Ice particle

Water droplet (supercooled)

(b)

Vapour sublimates around ice particle, which grows larger at the expense of the droplet (i.e. glaciation)

vapour flux

Droplet grows smaller by evaporation to maintain SVC in adjacent air. Droplet will eventually be completely evaporated unless growing ice crystal is precipitated first

Figure 8.2 *Schematic representations of the Bergeron-Findeisen process of hydrometeor growth.*

and the juxtaposition of droplets and crystals in the uppermost parts of the cloud. It also accounts for the success of cloud seeding with dry ice (solid carbon dioxide crystals) and silver iodide (freezing nuclei) in portions of cloud with an abundance of supercooled droplets (see case study in this chapter). However, ice crystals can grow large enough to overcome the terminal velocity only when the ice crystal concentration is small (for example, one crystal per thousand supercooled droplets) and when the cloud's liquid content is small, with cloud temperatures in the range −10 °C to −30 °C. Under these conditions, this large size (several millimetres across) will be rapidly attained in about 10–30 minutes. The process cannot be applied to hydrometeor growth in the tropics, where towering cumulus clouds frequently yield torrential rainfall without reaching the freezing level. Here, coalescence of coexisting warm and cold droplets is more important, possibly associated with the development of an adequate vapour pressure gradient between the warm droplet (high vapour pressure) and cold droplet (low vapour pressure).

Finally, accretion can also take place in clouds containing both supercooled water droplets and ice crystals, and occurs when ice crystals are swept aloft by updraughts into a zone of supercooled droplets with a concentration of less than 1.0 g m^{-3}. On striking the flake, the droplet freezes almost immediately on its surface to produce a rimed snowflake. If accretion is more pronounced in clouds with a liquid content well in excess of 1.0 g m^{-3}, then the frozen droplets can accumulate as rime ice to form soft, fragile graupel. Also, at −10 °C the droplets can freeze directly into hard, durable clear ice as ice pellets or hailstones, an integral feature of thunderstorms described at the end of this section (and see Chapter 16).

There are definite associations between the type of hydrometeor and the free or forced convective mechanism in operation and whether or not stability or instability exists (in the same way as cloud forms are controlled, as discussed earlier). For example, in stable conditions, hydrometeors are characterised by their small size, steady fall and a tendency to persist over a long period of time as drizzle, gentle rain or ice needles. These types of precipitation are particularly common at stable warm fronts (see Figure 14.6b) and with an updraught of air over gentle mountain slopes. The best indicator of stability is the deposition of freezing rain or glaze ice, since its formation depends on a fairly strong temperature inversion normally associated with a winter warm front. Here, temperatures are above 0 °C in a shallow zone of advected warm air (say between 1000 m and 2000 m) away from freezing conditions at the Earth's surface.

Unstable conditions lead to the formation of hydrometeors with a large size, intense fall and intermittent behaviour. They develop as large raindrops, hailstones or large snowflakes, and the resultant weather in buoyant free convection systems is dominated by severe, showery squalls. These types of precipitation are also experienced at unstable warm and cold fronts (see Figure 14.6a and 14.6c) and with the violent updraught of air along steeply sloping mountain flanks. The most outstanding effect of instability is the continued growth of cumulonimbus clouds, resulting in thunderstorms with the release of torrential rain and hailstones. This storm development is illustrated as a single cell in Figure 8.3, where the vigorous updraughts and downdraughts (reaching 1800 m min^{-1}) cause intense friction between the air and hydrometeors, which leads to a very high voltage build-up. The positive charges are concentrated in the upper parts of the cloud, since they have an affinity for the ice crystals, whereas the negative charges build up in the lower parts. The latter results in a positive charge at the Earth's surface, which reverses the fair weather gradient. When the negative–positive voltage potential difference becomes high enough then an electrical discharge (lightning flash) takes place within the cloud and from cloud to ground. The rapid expansion and contraction of intensely heated/cooled air along the discharge path produces the explosive sound called thunder, which reverberates or rumbles because of the time differences required for sound to reach the observer from various parts of the flash. The light from the flash is received almost instantaneously, whereas the slower sound waves travel only 1.6 km in 5 seconds, producing the time-lag between flash and accompanying thunder.

The violent updraughts and downdraughts in a thundercloud prevent ice pellets reaching the ground. They are carried aloft and grow by the accretion of supercooled droplets as rime or opaque ice in the coldest, higher parts, where the liquid water content is small, and glaze (clear ice) when abundant large droplets freeze rapidly. A pellet that passes through several parts of the

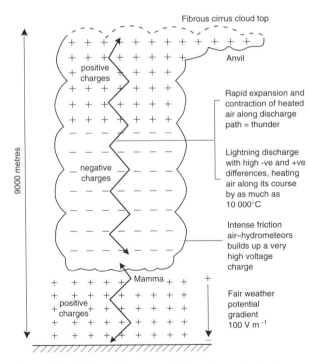

Figure 8.3 *Schematic representation of cumulonimbus cloud and thunderstorm development.*

thundercloud with varying liquid water content acquires alternate layers of clear ice (high content) and opaque ice (low content), forming a typical multi-layered hailstone growing as large as a golf ball. Eventually, its terminal velocity becomes so large that the updraughts cannot keep it aloft in suspension, and it falls rapidly to Earth, normally in the early stages of a cumulonimbus development. The actual circulation and life-cycle of a thunderstorm are discussed in Chapter 16 (Figure 16.4a).

CASE STUDY 8: Cloud seeding and precipitation modification

The case study at the end of Chapter 6 emphasised the urgent need to modify the hydrological cycle in attempts to solve water deficiency problems, especially in semi-arid lands. The discussion on deliberate vapour flux suppression reveals that small-scale, short-term success is possible, especially in terms of the construction of windbreaks and the application of chemical films in calm conditions. This case study now considers the techniques used in deliberately modifying the condensation–precipitation processes in an attempt to produce rainfall artificially.

Throughout history, human beings have sought to modify the rainfall regime, and primitive peoples employed witch doctors or medicine men to end drought and flood episodes in a variety of colourful ways. The first deliberate attempts to produce precipitation by artificial means were associated with General Dyrenforth in the USA between 1871 and 1872. He followed the 'explosion fallacy', whereby air parcels can be induced to rise in the troposphere with associated adiabatic cooling and dew point attainment (as discussed at the start of Chapter 7). Consequently, the General released oxygen-filled balloons into the air, which were exploded by his troops using mortar fire in the hope that the resultant perturbations would mirror free or forced convection (see Chapter 5). Indeed, this fallacy continued into the 1960s with the attempted correlation between wet periods and atomic bomb testing in the atmosphere of the South Pacific. However, it soon became apparent that even atomic bombs with the explosive power of the Hiroshima and Nagasaki explosions in 1945 release energy into the air equal to only an average summer thunderstorm in mid-latitudes (*viz.* 4×10^{12} joules).

Following this realisation, the explosion fallacy was abandoned and rainmaking attention became focused on cloud seeding as a viable, alternative technique (associated with the role of hygroscopic nuclei discussed earlier). Indeed, this technique was transformed by the laboratory projects of V. J. Schaefer (in 1946) and B. Vonnegut (1947), working in the US General Electric Research Laboratory. Schaefer pioneered cloud-seeding techniques when he used carbon dioxide crystals (i.e. dry ice), which he introduced into a freezer containing droplets cooled well below 0 °C (termed supercooled droplets). The introduction of a single crystal of dry ice (at about –80 °C) into a freezer of supercooled droplets (at about –23 °C) produced a trail of ice crystals (at a rate of 3×10^{10}), as it descended through the droplets. It was apparent that the juxtaposition of ice crystals and supercooled droplets provided great potential for the Bergeron–Findeisen process (explained earlier and illustrated in Figure 8.2).

Motivated by Schaeffer's discovery, Vonnegut searched for a chemical that had a crystallographic structure similar to that of ice, which could act as an artificial freezing nucleus and promote the phase change from vapour/liquid to a solid state. He subsequently found that a smoke of silver iodide particles produced numerous ice crystals in a cold box at temperatures considerably higher (–4 °C) than those required by dry ice. In practical terms, this meant that the essential freezing nuclei characteristics could operate successfully in warm clouds, at low elevations, where dry ice seeding fails (since the latter technique requires the upper reaches of deep cumulus congestus clouds (Figure 8.1). These laboratory discoveries were applied to the atmosphere in Project Cirrus between 1949 and 1951, when supercooled stratus clouds yielded the clearest response to seeding with dry ice. Initially, ground-based generators were used to seed with silver iodide and, in an experiment over New Mexico, success was claimed with increased rainfall over regions to the east of the state and especially over the Ohio river basin. However, seeding critics were not convinced and considered the observed favourable rainfall periodicity to be the result of natural variability rather than the cloud-seeding project.

Despite these reservations, Project Cirrus launched the USA and much of the world into the age of cloud seeding, referred to as the glory years of weather modification. For example, the Commonwealth Scientific and Industrial Research Organisation (CSIRO) in Australia conducted a silver iodide cloud-seeding programme over the New South Wales Tablelands during a five-year period in the 1960s. Cessna 310 aircraft (equipped with wing-tip smoke generators) were used to disperse 10^{17} freezing nuclei per hour in the upper reaches of clouds (at –5 °C) over randomly sampled parts of the tablelands. During this experimental period, the seeded areas were found to receive 19 per cent more rainfall than unseeded regions, although these promising results were not statistically significant. Furthermore, a major setback occurred in 1964, at the end of a six-year law suit following the Yuba City seeding disaster of 1955–56 in

northern California, USA. In December 1955, the Pacific Gas and Electric Company (PGE) employed North American Weather Consultants (NAWC) to seed clouds over the High Sierras to stimulate precipitation and replenish reservoirs for hydroelectric power demands. Following this seeding programme, NAWC claimed to have increased precipitation by 29 per cent and runoff by 123 million m^3. Unfortunately, these claims (without proven statistical significance) coincided with large-scale flooding at Yuba City, with 64 deaths and consequent damage worth US$65 million (at 1955/6 prices). The PGE and NAWC were assumed negligent and legally liable, and this litigation was continued until 1964 with the judge finally ruling that neither company was liable, although claims were successful against the State of California based on inadequate levee (flood bank) construction.

The problems of cloud seeding were apparent from this law suit, when the lack of control became the central issue. Even so, seeding programmes were extended to hail-suppression projects, particularly in the (then) USSR, Switzerland, Argentina, Italy, the USA and western Canada. Indeed, the Soviet Union reported a 50–100 per cent reduction in hail damage in the early 1970s, but attempts to replicate this success elsewhere have failed. Furthermore, seeding programmes have been used to reduce wind speeds in tropical cyclones, and Project Stormfury was initiated in 1962 between the US Weather Bureau and the US Navy. For nearly a decade, Stormfury performed a few tropical cyclone seeding experiments, and the best results were obtained from seeding Hurricane Debbie in 1969, with up to 30 per cent reduction in wind speed claimed. However, the project was abandoned in the late 1970s with no definitive results. Indeed at the present time, cloud seeding attracts little funding in the USA (compared with US$19 million per year invested in the mid-1970s). In fact, only Israel seriously considers this technique to be a viable solution to its water deficiency and in 1986, scientists at the Hebrew University in Jerusalem stated that rainmaking will become a 'way of life' in that country and will 'soon' double its annual rainfall totals. The evidence is awaited with considerable interest.

Keys topics for Part III

Atmospheric water vapour plays a vital role in the operation of atmospheric processes and systems at every scale:

1   The acquisition of water vapour by evaporation and evapotranspiration is initiated and sustained by atmospheric elements, particularly net radiation, vapour pressure deficit and wind speed (eddy diffusion).
2   Furthermore, environmental factors (namely water, soil and plant characteristics) control the effectiveness of the meteorological elements concerned and are modified deliberately by human activities in attempts to suppress vapour flux and conserve water.
3   The reverse condensation process is achieved when the air becomes saturated (i.e. reaches its dew-point temperature), mainly due to cooling mechanisms at or close to the Earth's surface and within the troposphere.
4   Surface cooling, due to heat loss by terrestrial (infrared) radiation, is responsible for the formation of dew, hoar frost and radiation fog, especially in valley/inversion situations. Surface cooling by air mass mixing produces advection fog, particularly when warm, humid airstreams pass over a colder surface at speeds of less than 8 m s^{-1}.
5   Fog dispersal and frost protection are operational human responses that attempt to minimise the inconvenience and costly disruptions to a wide range of economic and agricultural activities.
6   Condensation within the troposphere is related to the adiabatic cooling (DALR → SALR) of unstable parcels of air. Hydrometeors held in suspension produce layered or heaped cloud development and can grow to exceed the terminal velocity (especially by the Bergeron–Findeisen process) and fall by gravity as precipitation. This latter process is controlled artificially in cloud-seeding procedures using dry ice and silver iodide, with limited success.

Further reading for Part III

Cloud seeding potential over semi-arid lands. Russell Thompson, 1986. *Ambio*, XV(5), 276–281.
A general statement of cloud-seeding potential and techniques in water-deficient areas with a review of alternative techniques available to remedy water deficiency.

The Yuba City flood: a case study of weather modification litigation. Dean Mann. 1968. *Bulletin of the American Meteorological Society*, 49(7), 690–714.
An interesting account of a six-year law suit after cloud seeders were accused of losing control and inflicting death and destruction on a Californian city.

Human Impacts on Weather and Climate. William Cotton and Roger Pielke. 1995. Cambridge University Press.
A detailed, if technical, account of cloud-seeding principles and experiments applied to precipitation enhancement and hail suppression/tropical cyclone moderation. Part III emphasises the inadvertent changes due to aerosols and greenhouse gases discussed in Parts I and II of this book.

Climate, Water and Agriculture in the Tropics. I. J. Jackson. 1989. Longman Scientific & Technical.
A valuable and non-technical analysis of precipitation and evaporation processes in the tropics related to plant growth and agricultural potential.

Principles of Hydrology. R.C. Ward and M. Robinson. 1990. McGraw-Hill.
A detailed and non-technical account of all the atmospheric and Earth-surface pathways involved in the hydrological cycle.

 **Part IV**

The Primary Atmospheric Circulation: Global Pressure and Winds at the Earth's Surface and within the Troposphere

This section considers the primary circulation of the atmosphere and analyses the global pressure and wind characteristics and patterns that develop from the spatial variations of mass and energy balances at the surface and aloft, as discussed in Parts II and III. For example, within the humid tropics, positive net radiation values occur, with the major energy source in the form of latent heat flux from the tropical oceans and lush vegetation. The development of hot, unstable air parcels leads to vigorous free convection, and the rising moist air creates thermal low pressure (known as the inter-tropical convergence zone or ITCZ). The rising air parcels soon reach dew point, and the resultant saturated adiabatic lapse rate leads to deep cumiliform cloud formation and torrential rain. The associated release of latent heat has the potential to spawn active tropical cyclones/hurricanes (termed cyclogenesis) with highly curved inflowing winds. Meanwhile, high-pressure divergence aloft maintains this cyclonic circulation, with energy conversions contributing to vigorous air velocities.

9 The distribution of global surface pressure systems

Despite the complex and important linkages of surface pressure, upper air pressure, upper airflow and surface airflow in a three-dimensional atmosphere, it is convenient to start this section with an analysis of global surface pressure patterns (even though some of these systems, such as the subtropical anticyclones, have origins resulting from distinctive upper airflow and pressure regimes). Surface pressure differences then create surface airflows, which are studied in the next chapter and are coupled with upper air-pressure differences/airflow characteristics (see Chapter 11) in a complex cause-and-effect (i.e. so-called 'chicken and egg') relationship. This chapter covers:

- **the expression and representation of atmospheric pressure**
- **the basic mechanisms of low- and high-pressure systems**
- **the origin and characteristics of thermal and dynamic lows and highs**
- **global pressure distribution**
- **case study: the inter-tropical convergence zone**

Chapter 2 indicates that atmospheric pressure is measured by a mercury barometer, since it is represented (from the Earth's surface to the top of the atmosphere) by a column of mercury 760 mm high, which rises and falls in a tube (with a vacuum) immersed in a cistern full of mercury. An increased atmospheric pressure can support a larger column, and anticyclones cause the mercury to rise up the tube (with the reverse in cyclones). The height of the column is measured at fixed times to an accuracy of 0.1 mb, which is corrected for ambient air temperature since, as temperature increases, the density of mercury decreases and the barometer reads too high. These barometric pressure readings are also corrected for sea level, and simultaneous values, obtained from barometers distributed over a large area, are plotted on a map. Lines drawn through places with the same value of pressure (reduced to sea level) are called isobars, which reveal high-pressure centres (anticyclones) and their ridge extensions and low-pressure cells (cyclones or depressions) and their trough extensions (Figure 9.1). Slack-pressure areas between two adjacent high-pressure centres are known as cols. Indeed, the pressure field observed on an isobaric (or synoptic) chart resembles the contours on a topographic map sheet, where the same terminology is used (e.g. a topographic col or saddle is a high pass between two mountain peaks).

Box 9.1

The expression and representation of atmospheric pressure at the Earth's surface

1 Chapter 2 reveals that air has a definite weight (i.e. 1 m³ air weighs 113 g) and, consequently, the atmosphere exerts a force or pressure on the Earth's surface equal to 1.2 kg m⁻³.

1 Chapter 2 reveals that air has a definite weight (i.e. 1 m^3 air weighs 113 g) and, consequently, the atmosphere exerts a force or pressure on the Earth's surface equal to 1.2 kg m^{-3}.

2 Pressure decreases with height at a rate of 6 cm 1000 m^{-1} and varies spatially across the Earth, creating distinct areas of relatively low and high pressure.

3 Atmospheric pressure equals the force exerted by air on a unit area of a given surface. Its unit of measurement is therefore the unit of force divided by the unit area, which, in the metric system, is dyne cm^{-2}.

4 A dyne is the force necessary to accelerate a body 1 g in weight at a rate of 1 cm sec^{-1}. The numerical dyne value of atmospheric pressure is very large and approximates 10^6 dynes cm^{-2}.

5 Consequently, for convenience, a more practical unit of pressure has been adopted in meteorology, which is called a bar (which is exactly equal to 10^6 dynes cm^{-2}).

6 The bar is further divided into 1000 millibars (mb), so that 1 mb equals the force of 1000 dynes pressure upon an area of 1 cm^2. For example, the average pressure for the UK at sea level is 1,013,000 dynes cm^{-2} or 1.013 bars or 1013 mb (this latter value is easily represented on a surface pressure map, as in Figure 9.1).

The distribution of atmospheric pressure in Figure 9.1 reveals that anticyclones and cyclones/depressions can exist in juxtaposition. Furthermore, between these areas of high and low pressure, there exists a distinct gradient of pressure that is similar to the topographic gradients between hills and valleys, which are revealed by the contour spacing. Isobaric spacings indicate the prevailing pressure gradient, which is steeper when the isobars are packed close together (as at point X in Figure 9.1). Chapter 10 will confirm that airflow is initiated by the pressure gradient, since the winds blow from high pressure to low pressure at a velocity that is proportional to the isobaric spacings (see Figure 10.1).

Pressure differences observed on a surface isobaric/synoptic chart reflect an uneven distribution of the mass (weight) of air overlying that part of the Earth's surface. If atmospheric pressure at a station increases from, say, 980 mb to 1020 mb, as may happen in a day or two in the changeable weather regime of the British Isles (see Chapter 14), then this implies the addition of 4 per cent in the mass of overlying air. Conversely, a comparable pressure decrease means a withdrawal of that amount of air. Later in this chapter, it will be seen that in the development of high- and low-pressure systems, such apparently small pressure differences are in fact net changes resulting from very considerable horizontal transfers of air at different levels and in different directions. Furthermore, these horizontal airflows are linked

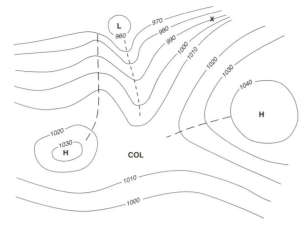

Ridge of high pressure
Trough of low pressure
Isobar (mb)

X   Point at which pressure gradient is steepest

Figure 9.1 A simplified isobaric/synoptic chart.

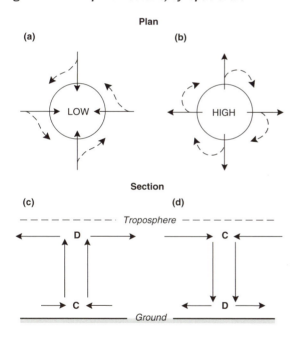

Figure 9.2 Basic airflow patterns in high- and low-pressure systems (Northern Hemisphere situation). In (a) and (b), the continuous arrows point towards lower pressure, with the broken arrows indicating the effect of the Earth's rotation (see Chapter 10) on the low-level flow. In (c) and (d), the lengths of the horizontal arrows are schematically proportional to the rates of inflow (convergence) and outflow (divergence).

by vertical motions (see Chapter 11) that are essentially upper air/ tropospheric systems and are shallow by comparison.

The basic mechanisms of high- and low-pressure systems

Figure 9.2 schematically illustrates the basic mechanisms of highs and lows in both plan and cross-sectional view. Later in this chapter, it will be seen that low pressure (i.e. a deficit of air) may be induced by various processes and that how ever initiated, it must be accompanied by an inward or convergent flow of air at low levels controlled by the pressure gradient (Figure 9.1 and above explanations). The equation of continuity (which restates the principle of conservation of mass) demands that a body of air that is compressed horizontally must expand vertically and rise. This is the meaning of convergence in the meteorological (rather than everyday) sense; it implies both horizontal and vertical motion (Figure 9.2c). Furthermore, at upper tropospheric levels, where further ascent is limited, the air column is effectively compressed vertically, which means that it spreads out horizontally: this is divergence (Figure 9.2c).

All low-pressure systems are, in fact, characterised by convergence below and divergence aloft, with rising air at mid-tropospheric levels in between (Figure 9.2c). The subsequent development of the system depends on the relative magnitude of surface inflow and upper air outflow, and with a persistent or deepening low, the outflow aloft must overcompensate for the inflow. If the reverse applies, the low 'fills'. Conversely, high pressure is characterised by divergence below and convergence aloft, and the vertical motion is one of descent or subsidence (Figure 9.2b and d). High pressure is maintained (and the anticyclone will intensify) only when upper-level inflow more than compensates for low-level outflow, otherwise the high will collapse. In reality, of course, the horizontal airflow depicted in Figure 9.2 is affected by the Earth's rotation and its deflection effect, which is known as the Coriolis force (see Chapter 10). This creates the characteristic directions of low-level circulation, which, in the Northern Hemisphere, are anticlockwise (cyclonic) with low-pressure systems and clockwise (anticyclonic) in the case of the high (i.e. the broken-line arrows in Figure 9.2a and b). It must be remembered that, in both these airflow systems, the sense of air circulation reverses with height.

The origin and characteristics of atmospheric pressure systems

This section examines the processes that generate the pressure/circulation systems discussed above (Figure 9.2) in terms of the required energy inputs and conversions involved, particularly in the thermal origins of low- and high-pressure systems. However, it will become clear that pressure systems also develop from dynamic or mechanical origins, where the forced ascent/descent of air (to create lows and highs) is more independent of thermal energy inputs.

A particularly simple example is the thermal or heat low, where the required energy input is thermal and is provided by an underlying hot surface. This occurs in equatorial latitudes and readily over islands and peninsulas in summer, when a considerable temperature contrast develops between heated land and adjacent cooler seas. Solar radiant energy (see Chapter 3) is converted into thermal energy at ground level, which in turn transforms into the potential energy embodied in the buoyant, expanding and rising air currents (Figure 9.2c), which then give rise to the kinetic energy (see Box 10.1) of the associated air circulations. In addition, condensation in the rising air parcels will release latent heat (see Chapter 5), which is a significant energy boost to any type of low. However, it is very important in the intensification of the thermal or thundery low (i.e. the storms that develop over Spain and drift northwards over France and the British Isles in summer), because of the warmth and high water vapour content of the air. When divergence aloft exceeds surface convergence, the lows intensify into destructive systems and, in tropical latitudes, they can develop into hurricanes or typhoons (see Chapter 15). However, these tropical lows are not solely thermal in origin, since they have more complex and composite origins, as will be discussed in Chapter 15.

Thermal, cold or glacial highs develop from the cooling of the land surface, particularly over Antarctica, Greenland and large continental interiors in the winter months of middle/high latitudes north of the Equator. Indeed, the highest mean sea level pressures recorded are those associated with the Eurasian or Siberian thermal high-pressure system (*c.* 1070 mb), which is also the most extensive. In response to severe radiational chilling (see Chapter 3), the lower layers shrink vertically and therefore spread horizontally to give the required surface divergence. Also, when coupled with vigorous convergence aloft, these systems intensify and, with the excess weight due to the concentration of cold air near the Earth's surface, this type is also classed as a cold anticyclone. This is essentially a shallow system with the overlying, coupled cyclonic circulation (see Chapter 11) often discernible at a height of only 2 km. Thermal highs can exist for long periods of time, like the infamous blocking highs, when the coupled upper-air systems are favourable (see Chapter 11 and Figure 11.8). Conversely, more transitory and temporary thermal highs (formed in polar air) often separate the travelling wave depressions of middle latitudes (see Chapter 14).

Dynamic low-pressure systems develop when air is forced to rise by frontal or orographic effects, which initiate the necessary surface convergence. Indeed, the wave depressions of middle latitudes develop along the Polar Front, which separates tropical and polar air masses (see Chapter 12) and provides the required dynamic or mechanical uplift. However, thermal energy conversions are still necessary, since the air masses represent zones where the initial solar energy input has been converted into differing air temperatures and hence densities. The close proximity of warm, less dense air and colder, more dense air at the Polar Front represents a situation high in potential energy. In fact, the development of the wave depression through a typical life-cycle (see Figure 14.4) is concerned mainly with the replacement of warm air by cold air at the surface and the conversion of this potential energy into the kinetic energy of a stronger cyclonic circulation. The frontal origin of the depression already presupposes the low-level convergence that is necessary to satisfy one of the basic requirements.

However, both theory and experience emphasise the need for this essential low-level convergence to be coincident with, and be overcompensated for by, high-level divergence. This need (plus the high-level convergence required for an intensifying surface high) can readily be supplied by a coupling to a favourable upper tropospheric airflow in middle latitudes. Chapter 11 will confirm this essential coupling, which is evident in well-developed Rossby waves (see Figure 11.8b). Here, the forward (downwind) limb is a favoured area for low-level cyclonic development (cyclogenesis), whereas the rear (upwind) limb favours low-level anticyclogenesis (see Figure 11.18). Dynamic cyclogenesis is also apparent in the orographic low or lee depression (or trough, when a closed circulation does not occur). When a broad, stable airstream moves over a high mountain range, a ridge of high pressure forms on the windward side and a low or trough characterises the leeward side. While this development can be regarded simply as the piling-up of air on the windward side and an air deficit on the leeward side, it is apparent that the air ascending the windward slopes also suffers vertical shrinking (and anticyclonic circulation). Conversely, air descending the leeward slopes experiences vertical stretching, which manifests itself as a cyclonic circulation. Consequently, a lee depression is often found on the southern (leeward) slopes of the Italian Alps and here is known as the Genoa Low. This development funnels alpine air down to the Mediterranean coast with velocity increases due to major relief gaps to create the Mistral or Bora winds (see Chapter 16).

Dynamic high-pressure systems include the orographic highs on windward slopes discussed above and, especially, the so-called warm anticyclones of subtropical latitudes, where the air is relatively warm at lower and middle tropospheric levels. Consequently, unlike cold highs discussed above, these are deep systems and are strongly evident at the 500 mb level (6000 m) and above. The excess weight is attributed to a high tropopause level compared with polar regions (see Chapter 2) and relatively cold air in the upper troposphere. Warm highs, of which the subtropical highs (like the Azores and Bermuda Highs) are the major examples, are relatively permanent systems, though they vary in position and intensity. This variation is distinctly seasonal due to the high-sun and low-sun migration of the subtropical jet stream (see Chapter 11), which is the main (dynamic) cause of the subtropical anticyclogenesis. Upper-air convergence characterises the equatorial side of this jet stream at 10,000–15,000 m which results in the descent and subsidence (i.e. vertical shrinking) of air and strong surface anticyclogenesis/divergence (see Case Study 11).

Global pressure distribution

The global pressure patterns represent a combination of the dynamic and thermal mechanisms discussed in the last section. Indeed, Figure 9.3 illustrates these pressure patterns in a schematic and generalised way, consisting of a series of latitudinal belts or cells of distinct high- and low-pressure regimes. The equatorial zone or trough is characterised by thermal low pressure and is associated with the inter-tropical convergence zone (ITCZ), which is analysed fully in the case study at the end of this chapter. The trough is a rather vaguely defined zone (Figure 9.4) that tends to move with the migrating overhead sun between the Tropics of Cancer (June) and Capricorn (January). However, in the northern summer (Figures 9.4 and 9.5), it merges with the

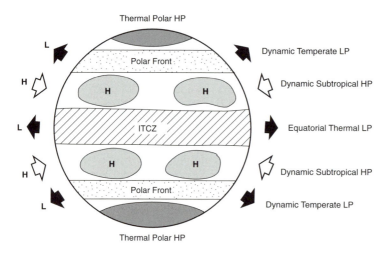

Figure 9.3 *Schematic representation of global pressure distribution.*

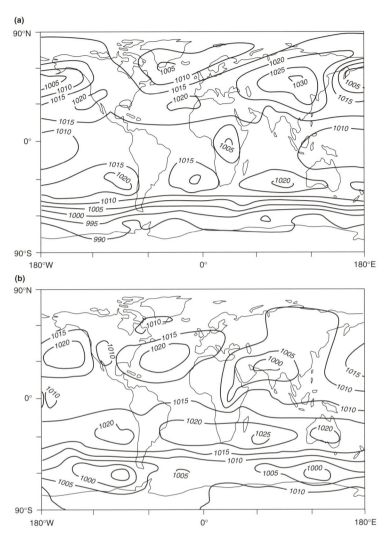

Figure 9.4 *Mean global pressure (mean sea level): (a) December to February and (b) June to August.*

huge thermal low that develops over southern Asia and the lesser one over the south-west of the USA.

The equatorial trough is flanked in both hemispheres by the dynamic subtropical anticyclones, which represent a broad belt of high pressure at around 30° latitude. As was discussed above, seasonal changes are associated with the low-sun–high-sun shift in the axis of the Subtropical Jet stream, which is the cause of this extensive anti-cyclogensis. High-sun migrations take the cells polewards and seawards, since thermal lows now occupy the land masses (shown schematically in Figure 9.5 and in Figure 9.4).

On the poleward side of these subtropical highs are located the so-called dynamic temperate/subpolar lows, which originate along the Polar Front, aided by favourable Rossby wave formations (discussed above and in Chapter 11). In winter in the Northern Hemisphere, these systems comprise the Icelandic Low of the North Atlantic and Aleutian Low of the North Pacific, but in summer they are ill-defined and tend to be absorbed by the continental thermal lows. In the Southern Hemisphere, however, the corresponding subpolar low is virtually continuous at about latitude 50° (Figure 9.4). These lows are purely statistical features since they have no real identity as such on individual synoptic charts and merely represent the high frequency of travelling low-pressure systems (wave depressions) in these latitudes (see Chapter 14).

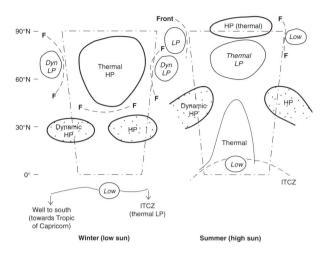

Figure 9.5 Schematic representation of seasonal changes in global pressure distribution.

Polewards of the subpolar lows lies the thermal/glacial high pressure of the polar regions, known as the North Polar and South Polar Highs. These shallow systems were discussed above and in the Northern Hemisphere, the high is very weak in winter compared with the higher pressures found over the snowbound lands of northern Canada and Siberia. Similarly, in Antarctica, the thermal high is most intense and persistent over the high ice sheets of the eastern zone, but it extends across the Southern Ocean in winter following the extension of the sea-ice cover (which virtually doubles the size of the Antarctic ice mass).

Case Study 9: The inter-tropical convergence zone

Circulation systems in the atmosphere over the humid tropics are primarily controlled by the well-known Hadley cell regime (Figure 9.6), which will be explained in Chapters 10 and 11. This thermally directed cell has two major low-level components (already discussed), namely equatorial (thermal) low-pressure convergence (the ITCZ) and subtropical (dynamic) high-pressure divergence at about 30° north and south, with associated subsidence and upper-air inversions (see Chapter 5 and Figures 9.3–9.6). These convergence and divergence zones are linked by a poleward airflow aloft and equatorward flow at the surface (the so-called trade winds), which will be discussed in the next chapter.

This case study concentrates on the characteristics and dynamics of the ITCZ, which dominates the weather and climate of the humid tropics. It develops on the equatorial limb of the Hadley cell, which, lying between the converging trade winds from both hemispheres, is a distinct surface low-pressure trough. This is nearly continuous around

the globe (when plotted on mean annual pressure charts, Figure 9.4), and can usually be traced upwards to the mid-troposphere. Originally (in the 1930s), this convergence zone was termed the inter-tropical front (ITF), since it was assumed to be thermodynamically similar to the Polar Front (see Chapters 12 and 14). Today, the ITF is recognised (as the inter-tropical discontinuity or ITD) over continental areas such as West Africa in summer (when hot, dry continental tropical air converges with less hot, humid maritime equatorial air). However, elsewhere, the term was replaced in 1945 by inter-tropical trough (ITT) and inter-tropical convergence zone (ITCZ). However, in practice it is difficult to differentiate between these two features (which depends on the angle at which the trades meet), so the term ITCZ is now commonly used to denote both circulations.

The ITCZ is characterised by great variability in both its structure and location. Structural contrasts are associated with the varying strength of the opposing trade winds, since weak convergence is

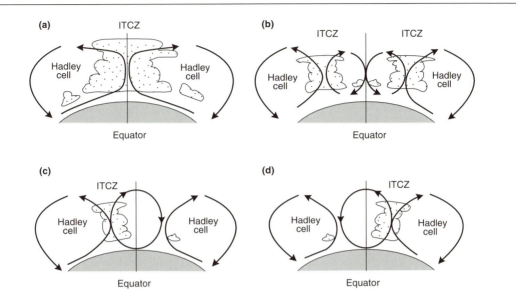

Figure 9.6 *The Hadley cell regime and the ITCZ (after Chang, 1972).*

associated with a poorly developed ITCZ with shallow cumulus clouds and little rainfall. Conversely, as the convergence intensifies, cumulonimbus clouds (see Chapter 8) become more common, with violent turbulence and torrential rain. Figure 9.6 illustrates the variety of forms evident in ITCZ development, with (a) representing the classical or textbook structure of a single mass of convergence over the Equator. However, subsidence–divergence can occur along the Equator (associated with radiational cooling at the cloud tops), and this results in the development of two new circulation cells–convergence zones in the adjacent tropical hemispheres (Figure 9.6b). Finally, the ITCZ can be represented by a single cell separating the Hadley circulations in both hemispheres (Figure 9.6c and d), and located either north or south of the Equator as a zone of heavy cloud and rain. Conversely, on the descending limb of the equatorial cell, pronounced subsidence occurs with a complete absence of heavy cloud development and rainfall.

It is now apparent that the ITCZ is not a region of continuously rising air with associated deep cumuliform clouds and heavy rain. Satellite observations have confirmed that a complete cloud cover does not occur in the region and that a series of well-developed cloud clusters (convergent zones) alternate with cloudless areas, with pronounced subsidence and divergence. Also, the convergent areas of cloud and rain have very varied structures, which range from linear to oval and circular formations. Furthermore, it is evident that these cloud clusters are commonly associated with westward-moving disturbances (see Chapter 15), where areas of convergence can grow or decay over a period of a few days.

The position of the ITCZ is generally assumed to vary seasonally in direct response to changes in location of maximum solar heating and the zone of seasonal maximum temperature (i.e. the so-called thermal equator). Consequently, in July, the ITCZ is likely to be located in the Northern Hemisphere, at around 25° N over hot continental South-east Asia and about 5–10° N over the comparatively cool Pacific and Atlantic Oceans (Figure 9.7a). In January, when the thermal equator moves towards the Tropic of Capricorn, the ITCZ is located mainly

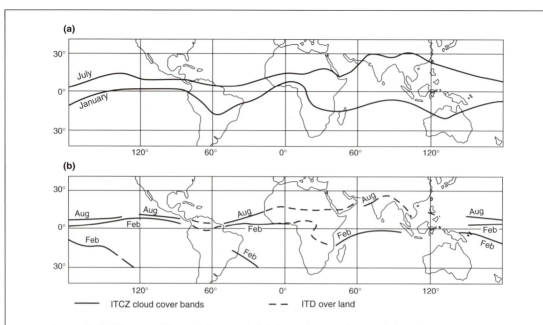

Figure 9.7 *Mean positions of the ITCZ: (a) August and February (after Henderson-Sellers and Robinson, 1987) and (b) July and January (after Barry and Chorley 1998).*

in the Southern Hemisphere, where (because of the smaller land masses and reduced continental heating) it is likely to be around 15° S over land and close to the Equator over the oceans. However, Figure 9.7 reveals that over the eastern Pacific Ocean (off Peru) and equatorial Atlantic Ocean (Gulf of Guinea), the ITCZ remains in the Northern Hemisphere, well to the north of the sun's zenithal latitude.

It is apparent from Figure 9.7b that the location of the ITCZ is more complex than the seasonal distributions discussed above. In the south-western parts of the Pacific and Atlantic Oceans, satellite observations reveal the presence of two semi-permanent convergence zones in February, both north and south of the Equator, related to long waves in the middle and upper troposphere (see Chapter 11). These zones are not found at this time in the eastern South Atlantic and South Pacific, owing to the occurrence of stabilising cold ocean currents (see Figure 12.2). Also, the convergence observed in the western South Pacific in February is now recognised as an important discontinuity. It is a zone of maximum cloudiness termed the South Pacific convergence zone (SPCZ), which extends from the eastern tip of Papua New Guinea to about 30° S, 120° W.

⑩ Surface airflow

The movement of air is a fundamental process in atmospheric systems, and horizontally moving air (wind) across the Earth's surface is one of the most obvious aspects of weather observed by the person in the street. It is less obvious to realise that wind is, in fact, the movement of air relative to the underlying surface (i.e. if both move at the same speed and in the same direction, then calm conditions prevail). Like any other body, air will move only in response to external forces, and horizontal air movement represents the resultant of up to four separate forces, which are examined below. This chapter covers:

- **the production and dissipation of kinetic energy (the mechanism creating air velocity)**
- **forces controlling wind direction and velocity, namely pressure gradient, the Earth's rotation, surface friction and the curvature of flow/centripetal force**
- **the control of vorticity and angular momentum on planetary airflow**
- **global winds and the general circulation**
- **case study: models of the global/general circulation**

Wind represents air in horizontal motion, and the principal cause of this movement is the equalisation of the horizontal differences in air pressure discussed in the last chapter. Indeed, it has been noted already that wind will blow from high pressure (an area of divergence) to low pressure (converging), and the direction in which this wind will blow depends on the actual hemisphere involved. For example, Buys Ballot's Law (1857) states that with one's back to the wind, the lowest pressure lies on one's left in the Northern Hemisphere but to the right in the Southern Hemisphere. It follows from this simple law that in the Northern Hemisphere, the winds blow in a clockwise circulation around an anticyclone but anticlockwise around a cyclone or depression (and *vice versa* in the Southern Hemisphere).

Kinetic energy

Winds blow primarily due to the conversion of solar energy (see Chapter 3) into the four main energy forms at the Earth's surface as indicated in Box 10.1, namely internal, geopotential, latent and kinetic energy (with the last form representing the energy for air movement). Kinetic energy represents only a small proportion of the atmospheric energy (i.e. 2 W m^{-2} or 0.95 per cent of insolation at the Earth's surface, see Chapter 3). However, this small percentage is sufficient to keep the atmosphere active enough to balance out latitudinally the unequal distribution of global energy (see Figure 3.3). Also, since the amount of kinetic energy is constant within the atmosphere, then it must be generated at the same rate at which it is dissipated.

Kinetic energy is generated in the atmosphere by convective systems (see Chapter 5) linking rising warm air with sinking cold air. In such conditions, the heavy and subsiding cold air lowers the system's centre of gravity and reduces the geopotential energy. This is now converted into other forms of energy, especially kinetic energy, which increases in proportion to the decrease in geopotential energy and causes the winds to blow. This conversion also occurs when warm air moves polewards and is chilled by conduction, sinks and lowers its centre of gravity, etc. This is one of the reasons (along with angular momentum changes, see Box 10.3) why excessive wind velocities characterise middle to high latitudes (e.g. the Roaring Forties, compared with the equatorial Doldrums explained later in this chapter). The atmosphere's kinetic energy equals 140 W hr m^{-2}, and the average rate of dissipation of this energy by friction (see point 6 in Box 10.1) has been estimated to be 2 W hr m^{-2}. Consequently, from these figures, it is apparent that the total kinetic energy of the atmosphere would be dissipated in 70 hours. However, the rate of energy reduction would decline as the available kinetic energy is reduced and should equal a dissipation rate of about 36 per cent per day, or 13 days to reduce the atmosphere's reserve of kinetic energy by 99 per cent.

In practice, atmospheric kinetic energy is not dissipated at an even rate over the Earth's surface, due to variations in friction over, for example, smooth polar ice sheets and oceans and more uneven upland and forested areas. In fact, the rate of dissipation of kinetic energy is proportional to the wind's velocity, and this is why so much of the

---

Box 10.1

Energy forms resulting from the conversion of solar energy

1   Internal energy represents the energy of molecular motion or heat.

2   Geopotential energy is increased when heated air expands and raises its centre of gravity, and is decreased with chilling and a lowering of its centre of gravity.

3   Total potential energy results from the combination of 1 and 2 above.

4   Latent energy represents the energy stored in water vapour during evaporation/evapotranspiration, which is released with condensation as heat.

5   Part of the atmosphere's total energy (1–4 above) is finally converted into kinetic energy, which results in air movement.

6   The atmosphere is constantly converting energy from one form to another; for example, kinetic energy is returned to heat following friction between the airflow and the Earth's surface and within the air molecules.

kinetic energy in deep cyclonic systems is converted by friction into heat. Indeed, the main sources of atmospheric kinetic energy are the gigantic overturnings of air at the Hadley cells (see Figure 9.6), which generate the powerful, westerly subtropical jet stream of both hemispheres (see Chapter 11), and the huge oscillations of air at the Rossby waves (see Chapter 11), with contrasting polar and tropical air masses (see Chapter12) in juxtaposition.

Forces controlling wind direction and velocity

Chapter 9 indicates that surface airflow is initiated by the pressure gradient resulting from the juxtaposition of high- and low-pressure systems. Winds were assumed to blow from high pressure (divergence) to low pressure (convergence) at a velocity that is proportional to the pressure gradient revealed by the isobaric spacing (e.g. the proposed stronger airflow at Point X in Figure 9.1, where the isobars are closest together). However, this section will show that even though pressure gradient force is an important consideration, a number of forces (other than simple air pressure differences) interact to control the direction and speed of winds.

Pressure gradient force

Inequality of pressure constitutes a first and most obvious reason for air movement. When pressure varies with distance there is said to be a pressure gradient, and this would be expected to exert a force urging the air from high to low pressure until the inequality is removed. The pressure gradient (Δp) is represented by the following equation:

$$\Delta p = \frac{dp}{dx} \tag{14}$$

where the change of pressure (p) with distance (x) is at right angles to the isobars (Figure 10.1).

A pressure gradient would appear to be a situation high in potential energy, analogous to that of a U-tube in which the level of liquid in one limb is artificially kept higher than that in the other. When the obstacle is removed, the liquid flows horizontally and eventually the levels become equal. On this reasoning alone, an isobaric/synoptic chart should indicate surface winds blowing at right angles to the isobars, which represent the pressure gradient (Figure 10.1). Furthermore, this pressure gradient exerts a force that is equal to its own magnitude. Hence, it would be expected that the steeper the pressure gradient (shown by the closest spacing of the isobars), the greater would be the impelling force and the stronger the surface winds (e.g. point X in

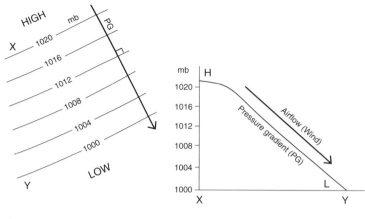

Figure 10.1 *Schematic representation of pressure gradient force.*

Figure 9.1). This relationship is also evident in Figure 10.2, where the steepest pressure gradients to the north-west of the British Isles are associated with the strongest winds (approaching gale force), compared with virtually calm conditions over the south-east (e.g. London), where the gradient is at its weakest. However, the wind directions in Figure 10.2 clearly are not at right angles to the pressure gradient (across the isobars) but are more nearly (but not quite) parallel to the isobaric trend. Obviously, while the pressure gradient is a primary cause of air movement, it cannot be the only one.

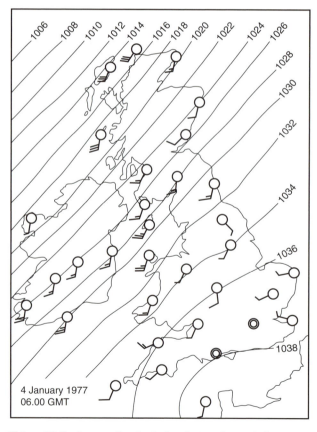

4 January 1977
06.00 GMT

Figure 10.2 *A synoptic chart showing surface winds over the British Isles.*

The Earth's rotation and the Coriolis force

It is apparent that the pressure gradient force is modified by the Earth's rotation, which produces a deflection force termed the Coriolis force (named after the French physicist G. G. Coriolis, 1792–1843), which accelerates moving bodies. Consequently, an object on or above the Earth's surface that is set in motion along a straight path (as seen by a fixed observer in space) will appear to an observer on the rotating Earth to be following a curved path. The effect is to deflect moving bodies to the right in the Northern Hemisphere and to the left in the Southern Hemisphere (which is Ferrel's Law of 1856) and is illustrated for the former hemisphere in Figure 10.3a.

Perhaps the clearest way to explain the Coriolis force is to represent the rotating Earth by a record player

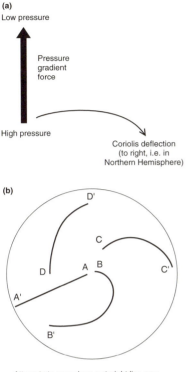

(a)
Low pressure

Pressure
gradient
force

High pressure

Coriolis deflection
(to right, i.e. in
Northern Hemisphere)

(b)

D'

C

D   A B   C'

A'

B'

Attempts to reproduce a straight line on a
rotating disc (B–D)

**Figure 10.3 *Schematic
representations of the Coriolis
force.***

turntable. Figure 10.3b illustrates this simple experiment
with four attempts to draw straight lines from the spindle
(pole) to the edge (Equator) on a paper disc fitted to the
turntable. A straight edge was placed across the cabinet
containing the record player so that it could be held in a fixed
position above the turntable. While the turntable was
stationary, there was (of course) no difficulty in moving a pen
along the straight edge to produce the straight line AA^1
(Figure 10.3b). When, however, the turntable was rotated
manually in a counter-clockwise direction (*viz.* to represent
the Northern Hemisphere), a line drawn along the straight
edge from B followed the highly curved path BB^1 (i.e. the
right-hand deflection proposed in Ferrel's Law). CC^1 and
DD^1 represent attempts to draw straight lines on the rotating
disc starting from points other than the centre. In each case,
the path traced on the disc indicates a weaker deflection to
the right of the original direction (as represented by the
straight edge). It is clear that nothing has happened to cause
this deflection except the rotation of the disc and the fact that
the points near the centre are moving more slowly than
points on the outer edge. This analogy can be applied to
movement across a rotating Earth, since a point at the
Equator travels at a speed of 470 m s^{-1}, compared with 230 m
s^{-1} at 60° and (like the turntable spindle) a rotation on the
spot at the poles.

The deflection effect due to the Earth's rotation is thus an
apparent force, yet one to be reckoned with. It can be given a magnitude, which (per
unit mass) depends on the angular velocity of spin about the vertical (Ω) and the
horizontal speed (V) of the air. Indeed, the magnitude of the Coriolis force (f) at any
point is expressed in the following equation:

$$f = 2\,\Omega\,V\,\sin\phi \tag{15}$$

where Ω is the constant rate of spin of the Earth, V is the variable speed of the
moving body and ϕ is the variable latitude. Consequently, air moving at 20 m s^{-1} is
subject to twice the Coriolis force exerted on air moving at 10 m s^{-1}. Also, the force
reaches a maximum at the North and South Poles and has zero value at the Equator
(this is why hurricanes do not develop in the equatorial zone, as explained in
Chapter 15).

The geostrophic balance

It is apparent that the initiating pressure gradient force (urging the air from high
pressure to low pressure) is opposed by the Coriolis force acting in the opposite

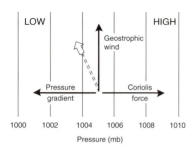

Figure 10.4 *Schematic representation of the geostrophic wind.*

direction. A state of equilibrium between these two forces results in a balanced airflow that blows parallel to the isobars, as illustrated in Figure 10.4. This supports Buys Ballot's Law, stated earlier, whereby, in the northern hemisphere, when an observer stands with his/her back to the wind, low pressure is to the left. The wind that blows in this balanced situation is known as the geostrophic wind (meaning 'Earth-turning'), where the velocity depends on the pressure gradient, shown by the isobaric spacings (as discussed in Chapter 9).

It must be emphasised here that the geostrophic condition is not always achieved in the atmosphere. Indeed, Figure 10.2 indicates that the surface airflow represented by this synoptic chart is obviously not balanced and is not blowing parallel to the isobars represented. Clearly, the surface winds blow from high to low pressure at an oblique angle to the isobars, which, in Figure 10.2, is remarkably constant over much of the British Isles. This is an example of ageostrophic flow, which, for near-surface conditions, is due to the intervention of yet another force associated with the frictional drag of the Earth's surface itself on the movement of air. Consequently, geostrophic winds are observed only above the zone of surface frictional retardation, usually above 600 m elevation. The jet streams (see Chapter 11) are good examples of geostrophic winds since they blow well above 6000 m.

Surface friction and ageostrophic airflow

The effect of surface friction is to slow down the movement of the lowest air layers. Since wind velocity is an important element in the Coriolis force (see equation 15), this side of the geostrophic balance is proportionally weakened and the more dominant pressure gradient is able to turn the wind direction somewhat towards the low-pressure (convergence) side. Thus, in Figure 10.4, the actual surface wind (shown by the broken arrow) represents the resultant ageostrophic flow, which tends to blow obliquely across the isobars (well-illustrated in Figure 10.2). This frictional effect (which is observed in a variable zone of the lower air layers, up to a height of 500–1000 m) depends on the actual surface roughness, which, for example, is greater over land than sea. Consequently, the oblique cross-isobar airflow averages 30° over rough, hilly land, compared with 10° over a smooth ocean.

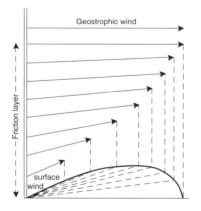

Figure 10.5 *The Ekman spiral: schematic representation of the change of wind speed and direction within the friction layer.*

Within the friction zone, wind speed reduces and the angle with the isobars increases as the Earth's surface is approached. This results in a wind spiral effect (Figure 10.5) first examined by, and named after, V. W. Ekman in 1905. The exact shape of this Ekman spiral varies with surface roughness, as the above land/sea example illustrated. Hence, the actual degree of spiralling in Figure 10.5 varies from a pronounced hook effect over rough land (where the surface wind may have only half the geostrophic value) to a more open curve over the sea (where the surface wind may have two-thirds of the geostrophic speed). Furthermore, at night over land, the surface wind may drop to a very low velocity and the angle of cross-isobar flow may increase to 40–50°.

Curved flow, centripetal force and the gradient wind

The discussion so far has concentrated on mostly straight isobars and geostrophic flow above the friction zone and ageostrophic (oblique) flow near the Earth's surface. However, when air moves in a highly curved path, an inward acceleration or centripetal force (c) becomes operative, which again destroys the geostrophic wind velocity. With curvature of flow in both low- and high-pressure systems, the centripetal force maintains the air in its curved path. This has a magnitude that is expressed as:

$$c = \frac{mV^2}{r} \qquad (16)$$

where m is the mass of moving air, V is its velocity and r is the radius of curvature.

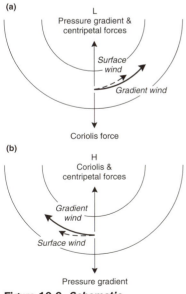

Figure 10.6 **Schematic representation of the gradient wind.**

Figure 10.6 illustrates the balance of the three forces involved, which determines the so-called gradient wind. In the low-pressure (cyclonic) example in (a), the inward acceleration or centripetal force weakens the outward-directed Coriolis force. This means that the wind velocity is now less than in the geostrophic flow and is termed subgeostrophic. Conversely, in high-pressure, anticyclonic systems (b), the centripetal force now weakens the outward-directed pressure gradient force and increases the wind velocity to supergeostrophic rates. However, the importance of this effect is reduced in anticyclones because the pressure gradient is indeed generally slack (weak airflow) and the radius of curvature (r) is large. However, in intense cyclonic circulations (especially deep depressions and hurricanes, see Chapters 14 and 15), where r is small or in curved flow at upper tropospheric levels, where velocities are large (especially pronounced Rossby waves, see Chapter 11), the combined effect may be considerable. In all these cases of strongly curved flow, the airflow that is represented by the isobar pattern is maintained by a balance of three forces (*viz.* pressure gradient, Coriolis and centripetal) and is referred to as the gradient wind. This wind blows in strongly curved circulations above the friction zone (500–1000 m). However, near the surface, frictional retardation is effective and the resultant airflow is now a response to four forces. These surface winds (Figure 10.6) are directed inwards across the isobars at the base of a low-pressure system and outwards at the base of a high.

The characteristics and dynamics of global winds and the general circulation

The principles and forces involved with surface airflow discussed so far will now be applied to the planetary or global circulation. However, before the circulation systems are discussed in detail, it is important to explain the variations in velocity and direction of flow that characterise surface airflow across the Earth's surface. These are associated mainly with the concepts of vorticity (an expression of the rotational property or spin of fluid particles) and angular momentum (the response of air velocity to changes in radial distance of the rotating point from the axis of rotation), as revealed in the following two boxes.

Box 10.2

The conservation of absolute vorticity

1 It involves a balance between the Coriolis force or planetary vorticity (PV) and the spin of the moving air mass or relative vorticity (RV).

2 Since the Coriolis force is zero at the Equator and maximum at the poles, as an air mass moves polewards its relative vorticity must decrease to conserve absolute vorticity (and *vice versa* as it moves equatorwards).

3 Figure 10.7 shows that this decrease in RV is compensating for the increased PV, which results in a negative spin and clockwise/anticyclonic curvature (Northern Hemisphere) towards the Equator.

4 As the air mass is redirected equatorwards, PV now decreases (zero Coriolis force at the Equator), and RV must increase to conserve absolute vorticity (AV). This results in a positive spin and an anticlockwise/cyclonic curvature (Northern Hemisphere) towards the North Pole, when stages 2 and 3 above are repeated.

5 This conservation accounts for the massive wave oscillations and airflow meanderings in middle latitudes (with amplitudes of 2000–3000 km). This is best observed in the flow of the Polar Front Jet stream (see Chapter 11) and the formation of well-developed Rossby waves, which dominate the troposphere in these latitudes and are responsible for cyclogenesis and anticyclogenesis (see Chapter 14).

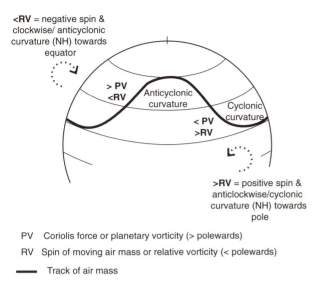

<RV = negative spin &
clockwise/ anticyclonic
curvature (NH) towards
equator

> PV
<RV Anticyclonic
curvature

Cyclonic
curvature

< PV
>RV

>RV = positive spin &
anticlockwise/cyclonic
curvature (NH) towards
pole

PV   Coriolis force or planetary vorticity (> polewards)

RV   Spin of moving air mass or relative vorticity (< polewards)

━━━   Track of air mass

Figure 10.7 *The conservation of absolute vorticity.*

Planetary/global winds are initiated in response to the global pressure patterns that develop across the Earth's surface, following the dynamic and thermal origins discussed in Chapter 9. Figure 10.9a illustrates (hypothetically) over a homogeneous globe the global wind patterns resulting from the pressure gradients imposed by these alternating high- and low-pressure zones (see Figure 9.3) developed between the Equator and both poles. Indeed, the solid arrows in this idealised figure indicate the initiating pressure gradient force as air moves in response to the alternating convergence and divergence. However, due to the sphere's rotation, the Coriolis force becomes readily apparent (broken arrows in Figure 10.9a), and the deflection to the right (Northern Hemisphere) and to the left (Southern Hemisphere) creates the well-known polar

---

Box 10.3

The conservation of angular momentum

1   The atmosphere rotates with the Earth and creates its own inertia around the Earth's axis: that is, the atmosphere possesses angular momentum (G).

2   The angular momentum of a body moving in a circle is proportional to its velocity and to the distance from the centre of the circle (which, in the atmosphere's case, is the Earth's axis).

3   G is the product of three factors, namely mass, velocity and radius (i.e. the radial distance of a rotation point from the axis of rotation), and is constant for any body moving around a fixed axis.

4   So, in the absence of any other force, any decrease of radius (i.e. a body moving towards the pole with mass constant) must result in an increase in velocity. Conversely, a body moving equatorwards, with an increase in radius, must experience a decrease in velocity (Figure 10.8).

5   A good example of G occurs with the spinning of an ice skater, since when the arms are pulled to the chest (with a decrease in the radius of rotation), the speed of spin must increase to conserve G, and a spectacular pirouette results. Conversely, to slow the velocity of spin, the arms are outstretched to increase the radius of rotation.

6   This conservation accounts for the strong winds in middle–high latitudes (e.g. the Roaring Forties) and the calm conditions in low latitudes (e.g. the Doldrums), which are explained in the next section.

When a body moves in a circle, its Angular Momentum (G) is proportional to

Mass x Velocity x Radius

i.e. if radius decreases and mass remains same, velocity must increase to conserve G and *vice versa*

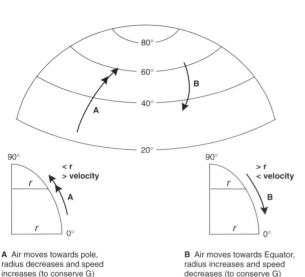

A Air moves towards pole, radius decreases and speed increases (to conserve G)

B Air moves towards Equator, radius increases and speed decreases (to conserve G)

Figure 10.8 *The conservation of angular momentum.*

Plate 10.1 *Satellite view of global pressure distribution, Meteosat 16/5/79. 1 = mid-latitude wave depressions (with distinct cold fronts); 2 = subtropical dynamic highs; 3 = ITCZ.*

easterlies, westerlies and trade winds between the poles and Equator in both hemispheres (Figure 10.9a). Figure 10.9b highlights the prevailing pressure/airflow relationships in the Northern Hemisphere in a simplistic way, assuming a globe with a uniform (ocean) surface. The north-east trades link the diverging air from the subtropical high with the converging air of the equatorial low-pressure trough, whereas the (south) westerlies are directed from the same divergence polewards to the subpolar low. This convergence zone also receives polar north-easterlies diverging outwards from the polar high. Figure 10.9b/c also includes a schematic cross-section from the surface up to 15 km in the troposphere that acts as a preview to the broad relationship between surface and upper airflow and the dominance of westerly winds at height (see Chapter 11).

The demerits of Figure 10.9b/c are associated with the impression given of a static pattern of pressure and wind belts. This is some way from the realities of constant variation due to distinct seasonal changes (see Figure 9.5), land and sea (see Figure 9.4) contrasts and the regular migration of pressure cells following the Earth's rotation. Also, the cellular pressure pattern in equilibrium with a complex pattern of airflow is also hypothetical, as was revealed in the ITCZ case study at the end of Chapter 9. However, despite these demerits, the global wind and pressure representations do provide a useful generalised framework of long-term associations (Plate 10.1).

Indeed, a sea journey from 60° N to 60° S does experience all of these so-called prevailing winds (Figure 10.9a) from time to time, although synoptic (i.e. secondary) circulations can sometimes dominate the areas concerned and completely reverse the expected airflow. However, a dramatic 'switch' of air circulation as a ship moves into a different, discrete, 'zone' is never experienced. Perhaps the most dramatic deviation from the 'expected' prevailing global airflow occurs in the so-called westerlies, especially in the Northern

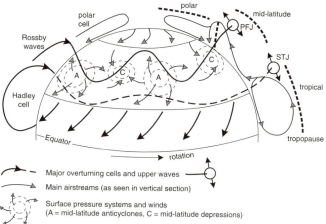

Figure 10.10 *The current model of the Northern Hemisphere primary circulation (after Hanwell, 1980).*

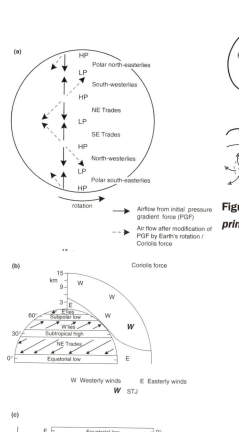

Figure 10.9 *Schematic representations of surface pressure and winds: (a) across the globe; (b) in the Northern Hemisphere; and (c) in the Southern Hemisphere.*

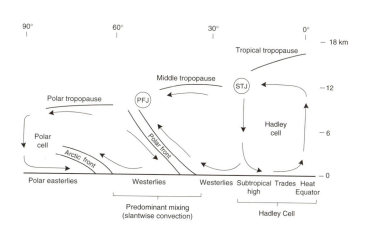

Figure 10.11 *Some features of the global circulation during a Northern Hemisphere winter (after Palmén and Newton, 1969).*

Hemisphere. Here, the conservation of both absolute vorticity and angular momentum (see Boxes 10.2 and 10.3) creates a meandering, high-velocity circulation. These distinctive oscillations create well-developed Rossby waves, which also initiate cyclogenesis and anticyclogenesis within the trough–ridge alternations (Figures 10.10 and 11.18). It is now generally accepted (Figure 10.10) that Polar cell/Hadley cells (Figure 10.11) do occur (with respective surface polar easterlies and trade winds) but that the westerlies are highly variable with distinct wave oscillations up to 16,000 m (see Chapter 11).

CASE STUDY 10: Models of the global/general circulation

The earliest ideas about the global circulation focused on the trade winds of the subtropics, which were interpreted by some sixteenth- and seventeenth-century philosophers as air left behind (due to its lightness) by the west–east movement of the Earth's surface and so constituted an easterly wind. Halley in 1686 mapped this thermal circulation as a single cell (Figure 10.12a), with air rising from the equatorial heat source causing a high-level pressure gradient that moved air polewards as a southerly wind. At the poles, the air descended over cold surfaces and returned to the Equator as a northerly surface wind. Halley was unaware of the importance of the Earth's rotation and explained the westward flow of the trades as yet another thermal effect, with air impelled towards a low-pressure zone under the westward-moving sun.

Almost half a century later, Hadley (1735) did incorporate the effect of the Earth's rotation and modified the simple Halley cell accordingly. Consequently (Figure 10.12b) Halley's upper poleward airflow became westerly, with the return surface flow now easterly. Even though Hadley ignored large areas of known surface westerlies, the notion of what has become known as the Hadley cell persists to this day (Figures 10.10 and 10.11), although it is now restricted to lower latitudes (in winter) between the Equator (ITCZ) and subtropics (dynamic anticyclogenesis), as

discussed in Chapter 9. During the second half of the nineteenth century, the concept of circulation cells became firmly accepted, and Ferrel (first in 1856 but more fully argued in 1889) was the earliest advocate of the three-cell model (Figure 10.12c). Ferrel maintained the Hadley cell concept but confined the descending limb to the subtropics, where the dominant subsidence (surface divergence) was due to the conservation of angular momentum (Box 10.3) aloft as an accumulation of poleward-moving air (convergence), having acquired maximum westerly momentum. Descent and low-level divergence drive the trades returning equatorwards (to maintain the thermal Hadley cell) and the surface westerlies of mid-latitudes flowing polewards.

These westerlies, in turn, form the low-level limb of a second (indirect) cell with rising air in subpolar latitudes. This eventually subsides over the poles to maintain a third cell, involving descending air in the polar high and surface easterlies (Figure 10.12c). The low-latitude and high-latitude cells are both thermal or direct, driven by the assumed temperature differences between the Equator and subtropics and between the polar ice caps and subpolar regions, respectively. Furthermore, the direction of surface airflow, given the Coriolis deflection, is consistent with the accepted (if idealised) circulation of Figure 10.9b. However, the middle (indirect) cell was assumed to be driven by

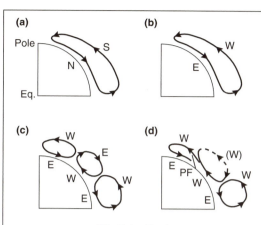

(a)

Pole

Eq.

S

N

(b)

W

E

(c)

W

E

E

W

W

E

(d)

W

E

PF

W

(W)

W

E

PF Polar Front

Figure 10.12 *Models of the global circulation, 1686–1941 (after Barry, 1969).*

the thermal cells to the north and south but was poorly explained. Indeed, the prevailing easterly airflow aloft in these cells was an erroneous (if logical) representation, since cirrus cloud movement reveals a westerly airflow at these altitudes.

The final elaboration of the three-cell model was associated with Rossby (1941), who retained the notion of thermally driven (direct) cells in low and high latitudes and regarded the middle cell as indirect or frictionally driven (Figure 10.12d). Rossby recognised that the circulation of this middle cell was westerly from the surface to the upper limb and explained this (illogical) reasoning by suggesting that the air was frictionally dragged in a westerly direction by the adjacent westerly air streams. The Polar Front (see Chapter 12) featured for the first time in circulation models (PF) as an important zone of convergence and cyclogenesis

(see Chapter 9). However, despite the success of this three-cell model, Rossby had abandoned this scheme by 1949, partly due to the difficulties attending the interpretation of the proposed middle cell. Also, the abandonment was partly a reaction to the fact that the cells alone were not capable of transporting surplus equatorial heat polewards.

It became apparent that this essential task was carried out by the large-scale horizontal eddies and oscillations in middle latitudes. These are the Rossby waves (discussed in the last section), which are responsible for migrating highs and lows, which shift deep streams of warm air polewards and cold air equatorwards (see Chapter 11). The whole middle-latitude circulation has been described as a 'slantwise' convective system, which is difficult to convey in a single simple diagram. However, its predominant mixing effect (in the winter circulation of the Northern Hemisphere) was included in a circulation model by Palmén and Newton (1969), which is illustrated in Figure 10.11. Both physical and numerical models support this view of the global circulation (see the 'dish-pan' simulations, explained in Chapter 11 and Figure 11.5), which is also illustrated more clearly in a current version of the general circulation (Figure 10.10). The predominant mixing by the Rossby waves is paramount in any general circulation model, especially when the amplitude of these waves is at a maximum. However, the next chapter will indicate that these wave patterns are sometimes poorly developed, which implies a minimal fulfilment of the circulation's basic role (see index cycle in Chapter 11 and Figure 11.6).

Upper troposphere pressure and wind systems

The previous two chapters in this section have already referred to the role of upper atmosphere pressure and wind systems in the development of their surface counterparts. Indeed, the introduction to Chapters 9 and 10 emphasised this complex cause-and-effect (viz. chicken and egg) relationship, where the coupling of surface systems with those aloft represents the essential three-dimensional atmospheric circulation. This chapter examines in detail the origin and characteristics of upper troposphere pressure patterns and indicates their influence on upper-air wind systems. It also emphasises the vital upper-air/surface linkages, especially regarding their control on weather regimes (including the monsoon of India and South-east Asia). This chapter covers:

- the representation of pressure in the upper troposphere
- the basic mechanisms of low- and high-pressure systems aloft
- the global distribution of upper-air pressure patterns, including the Rossby waves
- the controls and characteristics of upper-air wind systems, including jet streams
- the linkage of upper troposphere and surface circulation systems
- case study: upper-air pressure and wind patterns and the generation of surface weather systems

Tropospheric pressure patterns are represented on a map in a different way to the isobaric charts of surface pressure (see Figure 9.1). Indeed, because of temperature/pressure complications with increasing elevation it is more appropriate to depict the actual height of a selected pressure level by using so-called contour charts (usually drawn for 700, 500, 300 and 200 mb surfaces). For example, reference to the contours of a 500 mb surface indicates constant pressure heights around 5000–6000 m (or 500–600 decametres, dm) with the data acquired by radiosonde analyses of the upper troposphere. The heights of the actual pressure level aloft vary considerably over time and space; for example, the 500 mb surface varies between 6000 m and 4500 m (600–450 dm), with the highest levels being associated with the warmest conditions (and expanding air). Consequently, high heights correspond to high-pressure systems and low heights to low-pressure systems. This relationship is shown on Figure 11.1, both in plan form and as a vertical section, with large heights/high pressure in the warm expanding air (especially over the tropics or mid-latitude summer) and smaller heights/low pressure in the cold, contracting air over the polar regions or mid-latitude winter). Interestingly, the slope down from C (high pressure)

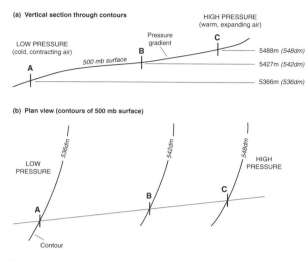

Figure 11.1 *The representation of upper atmosphere pressure patterns by contour maps.*

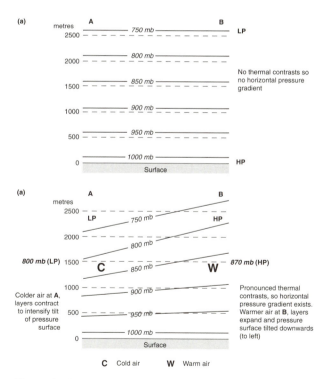

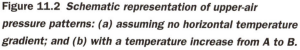

Figure 11.2 *Schematic representation of upper-air pressure patterns: (a) assuming no horizontal temperature gradient; and (b) with a temperature increase from A to B.*

to A (low pressure) represents the pressure gradient force (see Chapter 10), which motivates the upper airflow (as discussed later in this chapter).

Upper-air pressure systems

Figure 11.2 represents schematically a portion of the lowest 2500 m in the troposphere in the form of a cross-section along the line from point A (a polar latitude) to point B in the tropics. In (a), it is assumed that there is no pressure or temperature gradient at the Earth's surface and that the fall of temperature with height (the ELR of Chapter 5) will therefore be reached at uniform heights as shown by the contour lines, which are necessarily parallel with altitude. However, in (b), it is assumed that the air at B has been warmed (as indicated by a large W), while air at A has been cooled (large C). The air layers are now expanded in the warmed section and contracted in the chilled section, and the pressure surface is tilted from B down towards A. As a result, high pressure builds at B and low pressure forms at A, with a resultant pressure gradient at all levels except the surface (for example, at 1500 m the pressure at B is about 870 mb, compared with 800 mb at A).

Contour patterns thus mirror the distribution of mean temperature throughout the horizontal layer of atmosphere below the level specified. Consequently, the distribution of pressure globally throughout the troposphere will reflect the surface temperature regimes, particularly in response to the warm tropics and cold

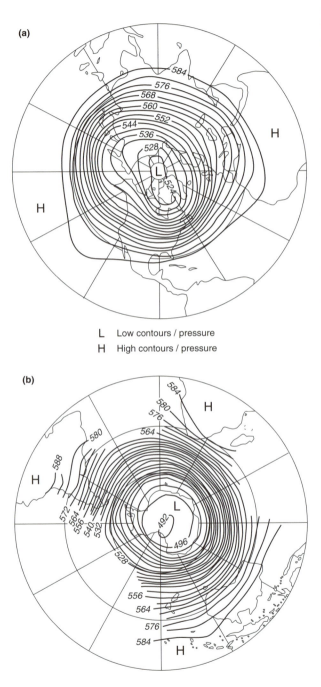

(a)

L   Low contours / pressure
H   High contours / pressure

(b)

Figure 11.3 *Annual mean 500 mb contours: (a) Northern Hemisphere and (b) Southern Hemisphere. Contour values are in decametres (dm).*

polar regions. Applying these considerations on a global scale, it can be seen that the cold polar ice-caps are represented by low contour (pressure) values. This simplistic distribution is clearly seen in Figure 11.3, where the tropical highs (H) and polar lows (L) are very obvious (with the lowest pressure recorded over the coldest Antarctic ice sheet, 492 dm compared with 524 dm over the Arctic basin). Although the Northern Hemisphere chart (Figure 11.3a) contains some interesting details (discussed below), the main impression is that of a concentric pattern of contours simply controlled by the temperature distribution between the Equator and the poles.

This impression is based on an annual averaging process, which, to a large extent, masks a rather different reality, which is revealed in seasonal contour charts for the Northern Hemisphere (Figure 11.4), especially for the winter season (February). At this time of the year, the concentric, symmetrical contours displayed by Figure 11.3a are not evident and, instead, an alternating pattern occurs of troughs (extensions of lower values/pressure) and ridges (extensions of higher values/pressure). These alternating troughs and ridges may number between three and six (with a preference for four), according to the prevailing situation (discussed later in the context of the so-called index cycle), and follow each other sluggishly in a west–east progression around the North and South Poles. They are also apparent in Figure 11.3, especially (a) representing the Northern Hemisphere, although they are not so clearly recognised south of the Equator in (b), as will be discussed later.

Rossby waves

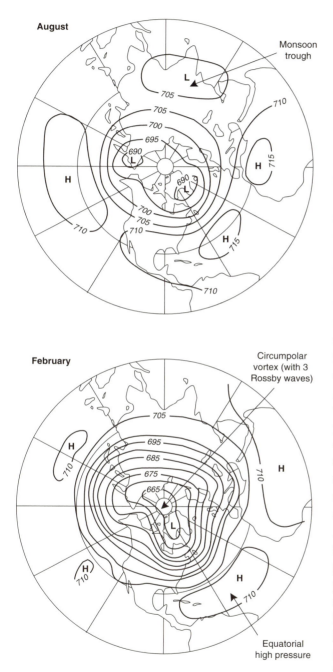

The preferred four locations of the upper trough axes are from the east Canadian Arctic to Florida, west Russia to the Black Sea, Alaska to the central Pacific and west Siberia to Burma, and the first three of these locations are evident in Figure 11.4. They constitute the long waves in the upper westerlies (up to 4000 km in length) and are generally known as Rossby waves, after the Swedish-American meteorologist who pioneered their investigation. The dominance of these large-scale wave motions/oscillations in the circulation of middle to high latitudes is related to the increase in Coriolis force polewards and the conservation of absolute vorticity (see Box 10.2 and Figure 10.7). This interpretation requires a mechanism to impose a poleward or equatorward component of motion on part of the general westerly flow (see Chapter 10) that, once initiated, maintains a pronounced meandering or wave-like oscillations. It appears that this mid- and upper tropospheric circulation is controlled by the topography of the underlying Earth. For example, the most distinct upper trough over eastern North America (Figures 11.3 and 11.4) lies in the lee of the Rocky Mountains, where peaks reach and exceed the 500 mb pressure surface. Such a formidable obstacle must disturb the westerly flow by a lee-troughing effect, which may initiate waves downstream. Large-scale seasonal pressure systems, resulting from thermal contrasts between continent and ocean (see Figures 9.4 and 9.5), could also influence long-wave development.

Figure 11.4 *Mean contours at the 3 km level, Northern Hemisphere.*

Box 11.1

The dish-pan representation of Rossby waves

1 A shallow circular dish containing water is placed on a turntable so that it can rotate about its centre.

2 The dish is heated at the rim to represent the Equator and cooled at the centre to represent the North or South Pole, thus simulating atmospheric conditions in one hemisphere.

3 A sprinkling of powder or filings on the water renders its motion visible, and a cine camera mounted above the centre (and rotating with the dish) records the evidence.

4 Slow rotation gives rise to a steady, symmetrical flow with highest velocities near the inner core.

5 However, at faster rotations, the Coriolis force becomes more apparent and the conservation of absolute vorticity (see Figure 10.7) produces an oscillating wave pattern (Figure 11.5).

6 This pattern resembles the Rossby waves in the upper westerlies (Figures 10.10 and 11.4) with high wind velocities concentrated in a narrow band along the zone of thermal contrast. This represents the Polar Front Jet Stream, explained in the next sub-section.

However, an alternative view is associated with the evidence from the well-known 'dish-pan' experiments, first conducted at the University of Chicago in 1951. There are obvious differences between a flat uniform dish and a hemispherical Earth's surface with pronounced topographical and thermal irregularities, but the dish-pan evidence is quite convincing in its simulation of Rossby waves. This suggests that these long waves might occur even above a uniform Earth, given the strong temperature gradient between the Equator and the poles. However, the lee-troughing effect discussed above (shown in Figure 11.3a) cannot be ignored, any more than their virtual absence from the corresponding chart of the Southern Hemisphere (Figure 11.3b). Here, ocean surfaces dominate and the high land (i.e. up to the 500 mb level) is confined to the narrow belt of Andean mountains. Consequently, the absence of well-developed Rossby waves (especially when compared with their northern counterparts) supports the significant role of the Earth's topography in their formation.

The index cycle

Synoptic analyses have revealed that the Rossby waves develop and decline in a distinct cycle, usually over a period of four to six weeks. This is known as the index cycle (Figure 11.6), during which the cross-latitude exchange of energy and vapour is discharged more vigorously at some times than others. Figure 11.6 indicates the four

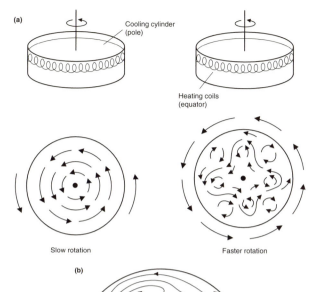

(a)

Cooling cylinder (pole)

Heating coils (equator)

Slow rotation

Faster rotation

(b)

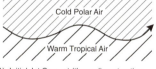

1) Initial Jet Core, at (thermal) contrasting juxtaposition of cold polar air and warm tropical air. Zonal airflow. High zonal index.

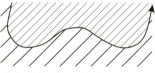

2) Oscillations increasing, due to conservation of absolute vorticity.

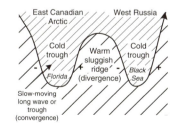

3) Great oscillations have developed, carrying polar air into middle and low latitudes and tropical air into middle and high latitudes. Jet stream meander very pronounced. Well-developed Rossby waves evident. Meridional airflow. Low zonal index.

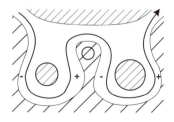

4) Extended troughs (waves) are cut off, leaving cells (cut-off lows) of cold polar air in the south and cells of warm air in the north. Low zonal index, soon to return to Stage 1 and high zonal index.

Figure 11.6 *Schematic representation of the stages in the index cycle.*

· Below average temperatures   + Above average temperatures   ⟶ Polar Front Jet Stream main stages in the cycle, starting with its initiation when the Rossby waves are poorly developed and the airflow roughly parallels the line of latitude. This represents a zonal airflow (see next sub-section) and a high zonal index type. As the wave oscillations increase through stages 2 and 3, the Rossby waves reach maximum amplitude and the airflow is now along the lines of longitude, representing a meridional airflow and a widely meandering Polar Front Jet Stream (see next sub-section), with the maximum poleward transport of energy and water vapour. This is called a low zonal index type. Finally (stage 4 in the cycle, Figure 11.6), as a consequence of severe distortion, the extended waves are cut off, leaving isolated bodies of cold air in warmer latitudes (the so-called cut-off lows, as in Figures 11.6 and 11.7) and detached bodies of warm air in colder latitudes (Figure 11.6). Experience has shown that the weather of mid- and high latitudes is largely dictated by the index cycle and the number/behaviour of Rossby waves. However, this is a complex relationship, since their movement is erratic, although generally it is in a west–east direction and much slower than the winds that blow through it. For example, some waves become stationary and some move east–west against the general westerly current. This can delay the termination of the index cycle and prolong the spells of extreme weather associated with blocking highs at the surface (during stage 3 of the cycle, as in Figure 11.8 over north-west Europe), which is discussed in a later section.

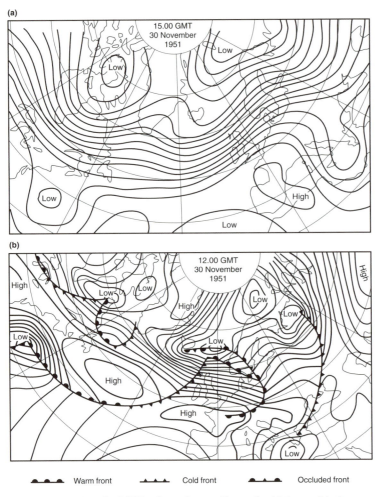

Figure 11.7 *A typical 500 mb contour pattern of a high zonal index and corresponding surface pressure distribution.*

Upper-air wind systems

Since the constant pressure surface contours are synonymous with isobars, it follows that the geostrophic wind aloft (i.e. free from surface friction, see Chapter

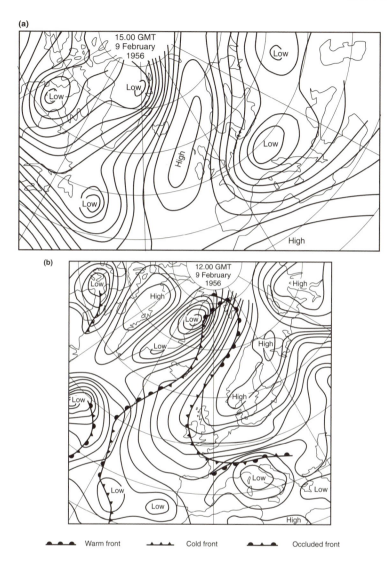

(a) 15.00 GMT 9 February 1956

(b) 12.00 GMT 9 February 1956

Warm front   Cold front   Occluded front

Figure 11.8 *A typical 500 mb contour pattern of a low zonal index and corresponding surface pressure distribution.*

10 and Figure 10.4) blows parallel to the contour lines. Indeed, since upper airflow is controlled only by the pressure gradient and Coriolis forces, then reference to Figure 11.9 will confirm an initial airflow (PGF) down the tilting pressure surface (from B to A), with a distinct Northern Hemisphere deflection by the Coriolis force into the paper. This airflow is termed the thermal wind since essentially it blows from expanding warm air to contracting cold air. It is also a geostrophic airflow that has an essentially westerly component, since the thermal wind blows with cool air (low pressure) to the left in the Northern Hemisphere and to the right in the Southern Hemisphere. Consequently, the poleward decrease of temperature in the troposphere causes a largely westerly airflow component. The main mechanisms of upper airflow are indicated in Figure 11.10 (based on the pressure surface contours represented in Figure 11.3). Here, in both hemispheres, the poleward pressure gradient force is balanced by the equatorward Coriolis force, and the thermal/geostrophic airflow aloft is predominantly westerly.

The obvious simplicity of airflow circulations aloft contrasts markedly with the complexities of surface airflow discussed in Chapter 10 and illustrated in Figure 10.9. Also, velocity variations are related simply to hemispherical thermal contrasts, where, for example, the colder air of Antarctica leads to a deeper low, steeper pressure gradient and more vigorous westerlies in the Southern Hemisphere than in the north (Figure 11.10). Figure 11.11 reveals that this pressure gradient is also much steeper in

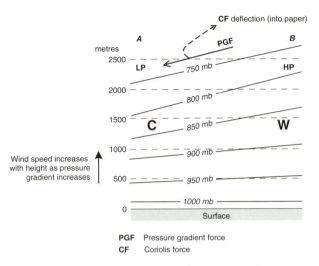

PGF Pressure gradient force
CF   Coriolis force

Figure 11.9 *Schematic representation of upper-air pressure patterns (Northern Hemisphere) and the resulting thermal wind.*

the winter season with the greater chilling of the North Pole basin and deeper low pressure. Consequently, the westerlies reach peak velocity at this time of the year, and it is also apparent that the greater polar chilling will intensify the Rossby waves. Figure 11.11 (February chart) confirms the increased amplitude of these waves, with a greater meandering of the westerlies and strong meridional flow (compared with the weaker, zonal airflow of the August chart).

A further complication is associated with the existence of high-velocity jet streams embodied in the westerly airflow at about latitudes 45° (Polar Front Jet) and 32° (Subtropical Jet). In fact, long before the introduction of upper-air charts, there was evidence to support the fact that wind velocities aloft far exceeded those at the surface. Indeed, in 1923, a balloon released at a fête in Hampshire, England, came to earth in Leipzig, Germany, four hours later, having covered the 917 km distance at an average speed of 64 m s^{-1}. Further evidence came from Zeppelins and other aircraft during both world wars, when strong winds at 5–6 km accounted for inaccurate bombing raids. In 1947, these high-velocity flows were christened jet streams by Rossby and his associates at the University of Chicago.

Today, the existence of jet streams is revealed clearly enough in contour charts of the upper troposphere, and analogous features have been reproduced in the dish-pan experiments described earlier (Figure 11.5). The 500 mb contour chart of Figure 11.7a confirms that very high wind speeds are in fact embedded in the Rossby waves, but the exact 'core' of the jet stream is only identified at higher levels. Indeed, two jet streams are clearly evident in the upper troposphere at the 300 mb level (9 km) and the 200 mb level (12 km), termed the Polar Front Jet Stream (PFJ) and the Subtropical Jet Stream (STJ), respectively. These jet streams are evident in Figure 11.12, which confirms their separate core locations in the Northern Hemisphere, although sometimes these two jets merge together as one high-velocity airflow. Both these major jet streams have counterparts in the Southern Hemisphere, although their meanderings are less pronounced (due to weaker Rossby wave development, discussed in the last section). Figure 11.13 confirms the location of the PFJ and the STJ over North America and Japan in December 1953, and the correlation of the PFJ with the Polar Front (see Chapter 12) is quite striking and provides a clue to jet stream creation (discussed below). Finally, an Easterly Tropical Jet Stream (ETJ) is also recognised over South-east Asia and southern India during the Northern Hemisphere

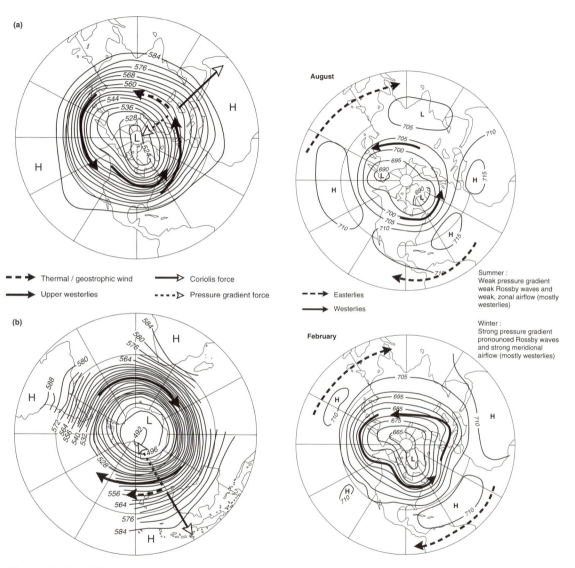

Figure 11.10 *500 mb contours: (a) Northern Hemisphere and (b) Southern Hemisphere, and idealised upper-air westerlies.*

Figure 11.11 *Idealised airflow at 3 km elevation in summer and winter (Northern Hemisphere).*

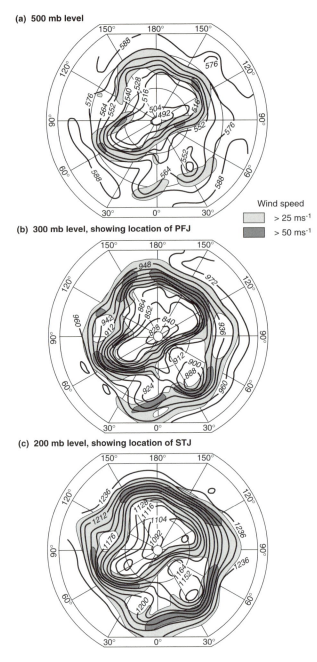

(a) 500 mb level

(b) 300 mb level, showing location of PFJ

Wind speed
▢ > 25 ms⁻¹
▨ > 50 ms⁻¹

(c) 200 mb level, showing location of STJ

Figure 11.12 *Location of PFJ and STJ 'cores' at the 300 mb/200 mb levels, respectively.*

summer (see Figure 11.16) and has a role to play as a 'scavenger' in the monsoon circulation (discussed in the case study at the end of this chapter).

The upper-air jet streams represent a zone of highly concentrated kinetic energy, which obviously derives from the energy transfer conversions highlighted in Chapter 10, Box 10.1, especially from enhanced potential and latent energy sources. At the Polar Front, the chilling of the dynamically rising air lowers its centre of gravity, converting total potential energy into kinetic energy. Also, the rising air is condensing as it cools and converting latent energy into the energy for air movement. Consequently, the PFJ accompanies the Polar Front, moving with it (although always displaced on the cold air side) and becoming involved in the distortions that occur with the development of frontal (wave) disturbances (see next case study and Chapter 14). The STJ is associated with energy conversions (and converging air from the Equator) along the poleward limb of the Hadley cell (see Chapter 9 and Figure 9.6). Here, the decreasing Earth radius releases potential energy, which is converted into kinetic energy and also influences the conservation of angular momentum, which increases wind speed to compensate for the decreased radius (see Box 10.3). Consequently, the STJ accompanies the Hadley cell/ITCZ movements (see Case Study 9 and Figure 9.7) and, for example, over Australia, migrates from the Tropic of Capricorn (in July) to about 45° S in January. The origin of the ETJ must be related to ITCZ energy conversions (potential/latent into kinetic), since it appears only during the high-sun/monsoon season.

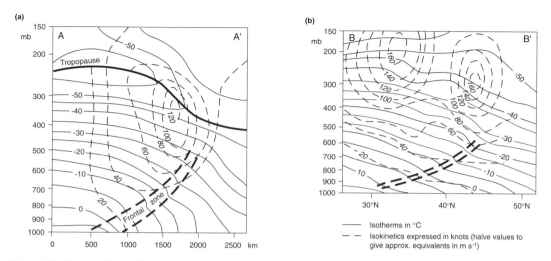

Figure 11.13 *Location of jet streams: (a) SW–NE section of the PFJ over North America, 19/12/53; and (b) NNE–SSW section of the PFJ (at c. 43° N) and the STJ (at c. 32° N) near Japan, 19/12/53 (after Sawyer, 1957).*

CASE STUDY 11: Upper-air pressure and wind patterns and the generation of surface weather systems

It is obvious that in a three-dimensional atmosphere, there is a strong (if complex) linkage between pressure and winds at the Earth's surface and similar features in the upper air. Indeed, this essential atmospheric coupling has already been emphasised in Part IV, especially in the origin of the dynamic anticyclone in the subtropics (see Chapter 9). In a simplistic and schematic way, this coupling was illustrated in Figure 9.2, when it became apparent that surface low pressure (convergence) was linked by rising air to upper high pressure (divergence). Conversely, surface high pressure (divergence) was linked by descending air to upper low pressure (convergence). Indeed, the intensification (or decline) of surface pressure systems demands excessive compensation aloft. For example, surface lows will intensify into very deep depressions or hurricanes only if the outflow aloft (in the high-pressure divergence) exceeds the surface inflow. If not, then the low will begin to fill

and eventually disappear. Similarly, surface anticyclones collapse when surface outflow exceeds the upper-air inflow, and these relationships account for the relatively temporary existence of travelling highs and lows in the westerlies (see Chapter 14).

Perhaps the most spectacular example of upper–surface air coupling is the generation of the semi-permanent subtropical anticyclones (see Chapter 9), which are responsible for the severe aridity of the world's great low-latitude deserts. Already in Chapter 9, the cause of this extensive anticyclogenesis was given as the STJ aloft. Indeed, the seasonal migration of these high-pressure cells (revealed in Figures 9.4 and 9.5) is clearly related to the movement of the STJ over the year. For example, the Australian example, referred to in the last section, indicated that the STJ migrates from the Tropic of Capricorn to about

45° S between July and January. At the same time, the centre of subtropical anticyclogenesis shifts in a similar way. It appears that the powerful convergence on the equatorial side of the STJ (i.e. the poleward side of the Hadley cell, see Figure 9.6) is responsible for subsidence over 12 km and surface divergence/dynamic anticyclogenesis.

The interrelationship between the (parent) STJ and the (offspring) subtropical anticyclone accounts for the distinctive mediterranean and monsoon circulations. The former situation affects regions around 35–40° north and south, which 'receive' the STJ and dynamic high in the high-sun season (Figure 11.14b). This leads to the hot, dry summers of the true mediterranean lands (such diverse areas as northern California, south-east Australia and south-west Africa). Furthermore, more poleward areas occasionally experience abnormal extensions of the STJ axis and dynamic high, which brings about extreme 'heatwave' and drought conditions in the UK and northern Europe (e.g. the infamous 1975–76 period). Similarly, the monsoon circulation is a most conspicuous feature and is also responsible for very pronounced seasonal changes of climate, partly due to the core changes of the STJ. Figure 11.14 illustrates the mean location and velocities of the STJ over the Northern Hemisphere in January and July. In the winter season, the jet core has returned to its subtropical 'home' (i.e. the Gulf of Mexico, North Africa and India), whereas in July it moves polewards over the Mediterranean basin and northern Asia.

The above seasonal STJ core changes over India represent the basic mechanics of the monsoon climate over the subcontinent. These changes are illustrated in Figures 11.15 and 11.16, which confirm that the dry winter season in India is associated with the STJ core aloft and anticyclogenesis/diverging offshore winds at the surface. Conversely, the wet summer season is associated with the movement of the STJ/anticyclone combination into northern Asia. This allows the ITCZ (see Case Study 9) to move into the subcontinent, with thermal low

(a) January

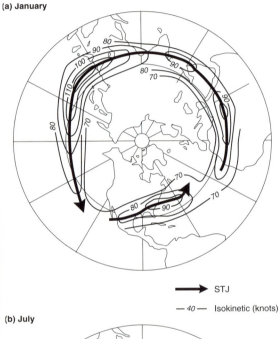

→ STJ

— *40* — Isokinetic (knots)

(b) July

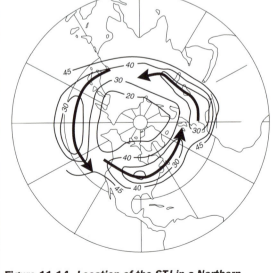

Figure 11.14 *Location of the STJ in a Northern Hemisphere winter and summer.*

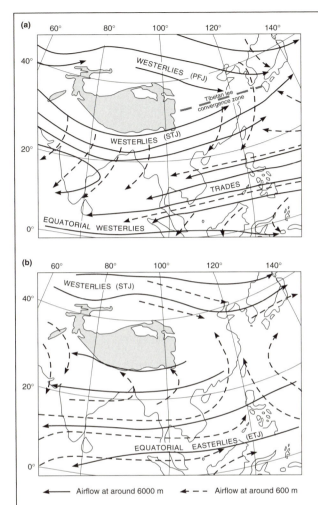

Figure 11.15 *Characteristic airflow of (a) winter and (b) summer monsoon circulations (after Barry and Chorley, 1998).*

pressure/converging onshore winds intensified by ETJ 'scavenging' aloft. These feed the free and forced convective systems and torrential rainfall, especially over the Himalayan foothills (where Cherrapunji has been known to receive 2500 cm rainfall in the handful of high-sun months). However, the south-west (wet) monsoon fails when the STJ/anticyclone do not move as far north as normal or when El Niño occurs (see Chapter 17).

There is no doubt that the strength and position of the Rossby waves influences surface weather systems.

Figure 11.17 represents this relationship in a schematic way, where troughs and ridges aloft have an obvious linkage with surface highs and lows, respectively. Furthermore, since the upper troughs represent equatorward outbreaks of cold polar air and upper ridges polar outbreaks of warm tropical air (see Figure 11.6), then the Rossby waves are also linked to extreme heatwaves and cold spells, in the mid-latitudes especially. Unfortunately though, the actual atmospheric couplings are not as simple as schematic interpretations suggest. This is confirmed in Figure 11.8 where comparisons of (a) and (b) do show the basic coupling idealised in Figure 11.17. For example, the 'blocking high' at the surface over Scandinavia does more or less couple with the Rossby trough and cut-off lows at the 500 mb level. However, close examination shows that the surface anticyclone is to the west of the main upper trough axis; in fact it is located more on the upper ridge (high)/trough (low) boundary. This is illustrated schematically in Figure 11.18, which indicates that the surface cyclogenesis C is located on the advancing/eastern edge of the upper trough, whereas surface anticyclogenesis A is located on the advancing or eastern edge of the upper ridge.

It is apparent that the PFJ has a role to play in the above 'off-centre' coupling, and Chapter 14 will reveal that contrasting zones of convergence and divergence along the PFJ are responsible for surface anticyclogenesis and cyclogenesis, respectively. Indeed, PFJ divergence is generally known as 'scavenging', which was observed above in the south-west monsoon intensification with the ETJ. The PFJ also steers wave depressions along its meandering course, and it also accounts for below-average

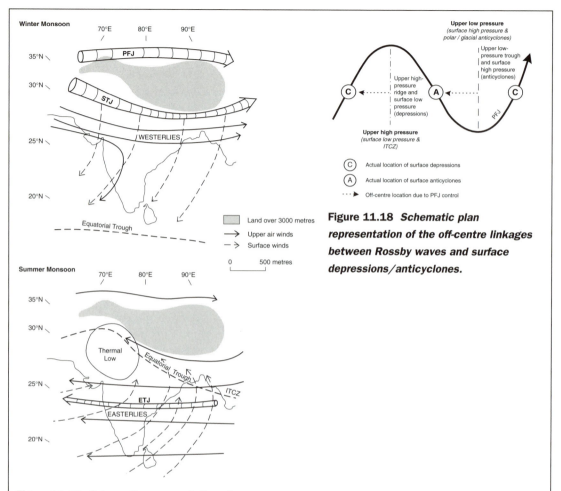

Figure 11.18 *Schematic plan representation of the off-centre linkages between Rossby waves and surface depressions/anticyclones.*

Figure 11.16 *Schematic representation of monsoon circulations.*

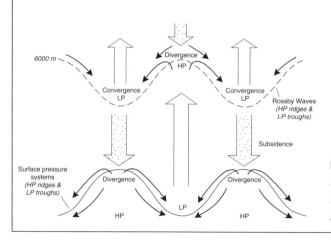

Figure 11.17 *Schematic cross-sectional representation of the coupling of upper-air Rossby waves and surface pressure systems.*

temperatures when it has a north-westerly trajectory and well-above average temperatures with a south-westerly jet trajectory. The high-velocity tail winds (flying east) and head winds (flying west) also affect the consumption of aviation fuel in jumbo jet aircraft and play havoc with timetabled arrival times (with delays flying westwards and early arrivals flying eastwards) when the core of a particularly strong PFJ corresponds with major routings.

Key topics for Part IV

The primary atmospheric circulation represents the response of global pressure and wind patterns to the spatial variations of mass and energy balances discussed earlier.

1 Pressure systems develop across the Earth's surface due to thermal and dynamic processes. These develop thermal lows in the tropics (ITCZ), thermal highs in the polar regions (glacial anticyclones), dynamic lows along the Polar Front interactive zone and dynamic highs in the subtropics (due to jet stream convergence aloft).

2 Winds are initiated by the pressure gradient developing from the juxtaposition of high- and low-pressure systems. This flow (from diverging highs to converging lows) is proportional to the prevailing pressure gradient but is modified by the Earth's rotation (the Coriolis force). When these two forces are balanced, the air flows parallel to the isobars as a geostrophic wind. However, surface friction reduces the velocity of the wind, which partially destroys the Coriolis force. The resultant flow is at an oblique angle across the isobars towards the low pressure convergence.

3 Planetary surface winds are controlled by the distribution of global pressure patterns, associated with thermal and dynamic orgins. The initiating pressure gradient force and counteracting Coriolis force create the well-known polar easterlies, westerlies and trade winds in both hemispheres. Furthermore, the high-velocity, meandering Westerlies owe their basic characteristics to the conservation of angular momentum and absolute vorticity.

4 Upper troposphere pressure patterns are relatively simple, with low pressure over polar regions (cold, contracting air) and high pressure over the tropics (warm, expanding air). However, due to the Earth's rotation (as was depicted in the dish-pan experiment), the pressure patterns are deformed into huge, long (Rossby) waves of alternating low-pressure troughs and high-pressure ridges. These wax and wane over time, as revealed in the index cycle.

5 Upper-air wind systems are mainly westerlies, which represent the thermal wind blowing from expanding warm air (highs) to the contracting cold air (lows). They blow in huge wave-like oscillations (following Rossby waves), especially in the Northern Hemisphere during the winter season. Jet streams are embodied in the westerly air flow, representing zones of highly concentrated kinetic energy in the vicinity of the Polar Front (PFJ) and subtropics (STJ).

6 Upper air–surface coupling of pressure systems is necessary to maintain the

essential three-dimensional linkage in the troposphere. For example, surface lows will intensify only into very deep depressions or hurricanes if outflow aloft (in high-pressure divergence) exceeds the surface inflow (and *vice versa* for intensifying surface highs). The monsoon circulation is an excellent example of the coupling of jet stream divergence or scavenging aloft (by the easterly jet), which favours surface thermal low (ITCZ) intensification.

Further reading for Part IV

The further reading for this section excludes the detailed coverage of the primary atmospheric circulation in the mainstream meteorology/atmosphere textbooks listed in the bibliography. Instead, 'classic' references are included here that, at the time, were 'milestones' in the development of our knowledge of global pressure and winds, at the Earth's surface and in the upper air.

The westerlies. F. K. Hare. 1960. *Geographical Review*, 50, 345–367.
A concise and non-technical account of circulation patterns in mid-latitudes, including their geographical extent and vertical structure. The Rossby waves and index cycle are also examined in the light of the 'discredited' three-cell circulation model.

Atmospheric Circulation Systems and Climates. J. H. Chang. 1972. Oriental Publishing Company.
A very detailed and comprehensive coverage of all the major circulation zones, namely the trades, ITCZ, westerlies, monsoon systems and polar regimes.

The geographer and the atmosphere. P. R. Crowe. 1965. *Transactions of the Institute of British Geographers*, 36, 1–19.
A probing, non-technical discussion of the general circulation that is both readable and thought-provoking. It asked some 'brutal questions' about the validity of Ferrel's Law and the geostrophic wind and attempted a 'new' interpretation of trade wind flow.

Part V Secondary and Tertiary Circulations: Synoptic Situations and Local Airflows

Part IV outlined the planetary or primary circulation systems, which act as a global 'framework' for the operation of secondary (synoptic) and tertiary (local) circulations, which are superimposed upon each other. Part V emphasises this superimposition of the airflows and examines the development and characteristics of the secondary systems (namely air masses and weather disturbances in the mid-latitudes and tropics) and tertiary airflows, especially local circulations due to topographic differences and urbanisation. The superimposition of atmospheric circulations and systems at every scale (from the planetary to the synoptic and local situations), determines the prevailing weather conditions operating at any place at any given time. It is expressed vividly by the following two quotations:

First, after Jonathan Swift – 'Great fleas have little fleas upon their backs to bite 'em, and little fleas have lesser fleas and so ad infinitum.'

Second, after L.F. Richardson – 'Big whirls have little whirls that feed on their velocity, and little whirls have lesser whirls, and so on to viscosity.'

⓬ Air masses and fronts

It is generally accepted that over certain large regions (e.g. Antarctica or the Sahara), the atmosphere can develop quite uniform air mass characteristics that reflect the thermal and moisture regimes of these contrasting environments. Furthermore, due to the alternation of travelling high- and low-pressure systems (see Chapters 9, 13 and 14), these air masses can move away from their distinctive source regions and, as airstreams, can invade neighbouring areas. Consequently, these invasions initiate secondary airflows, which can reverse the (expected) primary airflow and produce changeable weather regimes over a relatively short period of time. For example, the passage of a cold front over London, England, or Melbourne, Australia (where it is known as the 'Southerly Buster') can alternate a warm tropical airstream with a cold, polar airstream within a matter of hours. This chapter covers:

- air mass genesis and identification
- air mass properties in the source regions and modifications due to airstream invasion
- classification of air mass types
- air mass boundaries and frontogenesis
- case study: air masses over Europe, North America and Australia

An air mass is a body of atmosphere with dimensions of hundreds of thousands or, indeed, a few million square kilometres, as is the case over East Antarctica. Furthermore, an air mass displays little or no horizontal variation in any of its properties, especially temperature and humidity. Air masses move away from their distinctive source regions to become airstreams, when their recognisable source properties (especially as tropical or polar entities) are carried away by airflows motivated by the prevailing pressure systems (see Chapters 9 and 10) and, of course, these properties are modified en route.

Air mass genesis and identification

Air masses are described in terms of their basic temperature and humidity properties and the important derived property of stability (see Chapter 5), which depends on the distribution of temperature and humidity in the vertical. Thus, different air masses may be characterised as cold, moist and unstable or warm, dry and stable (or various other combinations of these properties), with the terms referring generally to the lower layers of the air mass concerned. Since the atmosphere is both heated and derives its moisture from the underlying surface, it follows that air masses inherit their basic properties by long-term residence (preferably over a month or more) over surfaces with uniform thermal and moisture characteristics, which are known as source regions.

Clearly, the essential requirement of a source region, apart from the homogeneity of surface, is a persistent anticyclonic regime, which alone allows sufficiently long stagnation of an air mass for the slow acquisition of thermal and humidity properties from surface to atmosphere. The major source regions are therefore the permanent or semi-permanent high-pressure regions of the global circulation (see Chapter 9 and Figures 9.3, 9.4 and 9.5). These include the polar (thermal) highs over the Arctic basin and Antarctica, the subtropical (dynamic) highs over the warm seas (Azores) and arid lands (Sahara) of similar low latitudes and the thermal highs that develop over mid-latitude continental interiors (Russia) when they are mantled by winter snow. The net outflow or divergence from the base of these long-term anticyclones ensures a persistent spread of associated air masses towards regions of lower pressure (see Chapter 10). From the point of view of air mass control, there are 'donor' regions and 'receptor' regions and it is in the latter that dramatic airflow changes and frontal activity are most pronounced.

Box 12.1

Identification of air masses

1 To be reliable indicators, the primary properties of the air mass should possess two essential characteristics, i.e. they should be representative and conservative.

2 Representative properties are not influenced by local conditions and must be typical of the entire mass of air rather than a small portion of it.

3 Temperature and humidity near the Earth's surface are not always reliable indicators, since they are so easily influenced by minor topographic or thermal differences at the surface itself. The vertical distribution of properties is a more meaningful indicator, especially the stability tendencies of the air mass, obtained from radiosonde ascents.

4 Conservative properties are not affected by changes in any other element. Specific humidity (i.e. the weight of water vapour per unit weight of air) and potential equivalent temperature (i.e. taking into account both DALR and SALR changes, see Chapter 5) are the most useful and conservative measurements available.

5 Secondary identification properties are more qualitative and include cloud type, type of hydrometeor and wind structure, especially as representations of stable and unstable conditions (see Chapter 8).

Air mass properties in the source regions and modifications due to airstream invasion

In the source region, air mass properties are quite obvious since they are simply related to the local environmental (thermal and moisture) conditions. For example, polar air (or air stagnating over cold mid-latitude continental interiors in winter) is very cold and stable since it is being chilled from the snow-covered surface with pronounced surface inversions (see Chapter 5 and Figure 5.3a). Conversely, tropical air is hot and unstable (up to the trade wind inversion, Figure 5.3b) since it is heated from below with a marked steepening of the ELR (see Figure 5.1). Similarly, air stagnating over a large land mass will be dry (continental characteristics) compared with the far more humid (maritime) air masses developing over the oceans.

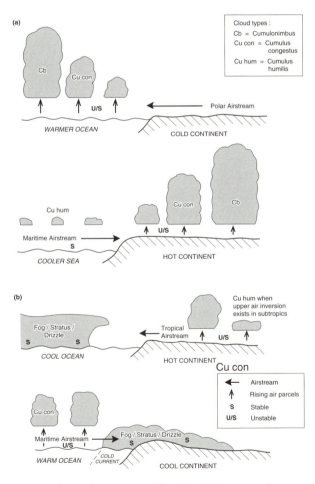

Figure 12.1 *Airstream modification: (a) passage from cold to warm surfaces; and (b) passage from warm to cold surfaces.*

As already mentioned, air mass properties are significantly modified by the passage of the airstream over surfaces that are different in character from the source region. Since all airstreams originate in anticyclonic regimes, they tend to be initially stable with a well-marked inversion due to upper-air subsidence or radiational cooling at the surface (see Figures 5.1 and 5.3). However, as a cold airstream moves equatorwards (or cool maritime air blows across a very warm land mass, Figure 12.1a) it will be heated from below with lapse rate steepening (see Figure 5.1) and will eventually become unstable in its lower layers. If its movement is over water, it will also acquire moisture by evaporation at an increasing rate since its temperature is also rising (see Table 7.1), with developing cumulus congestus cloud and heavy showers (especially during the daytime in the land mass scenario), as illustrated in Figure 12.1a.

On the other hand, when a warm air mass moves polewards (especially from a hot continent on to a cool ocean) or when a warm, moist airstream blows across a cold current (Figure 12.1b), then it is progressively cooled from below, with a strong temperature inversion developing in the surface layers, to induce air stability. The resultant chilling below dew point favours the formation of advection fog (see Chapter 7 and Plate 7.2) or low stratus cloud and fine drizzle (see Chapter 8). In terms of humidity changes, continental airstreams with a maritime trajectory will tend to become more saturated, especially as it moves from a tropical source region (e.g. Saharan air moving across the Mediterranean Sea in summer). Conversely, maritime airstreams with a continental track will retain their original moisture capacity unless they release water vapour in condensation, especially associated with forced/orographic convection. Consequently, the weather associated with a particular air mass in a 'receptor' region may be very different from that in its source region, and air mass identity is often destroyed due to its complex history.

Box 12.2

Geographical location and characteristics of main air mass source regions

1 High Arctic, Antarctica and excessively chilled northern mid-latitude continents in winter (Russia, Siberia, North America): very cold, dry and stable air mass with temperature inversions common = Polar Continental (Pc) type.

2 High-latitude oceans (Southern Ocean and north-west Atlantic), representing modified Pc airstreams with an ocean trajectory: cool, moist and unstable air mass with lapse rate steepening common = Polar Maritime (Pm) type.

3 Major subtropical deserts (Sahara, Australia) and excessively heated northern mid-latitude continents in summer (central USA): very hot, dry and unstable in summer; cooler, dry but more stable in winter = Tropical Continental (Tc) type.

4 Extensive subtropical oceans (Azores, Bermuda, Mauritius, Fiji): very warm, moist and unstable in west (over warm ocean currents, as in the Gulf of Mexico and the South China Sea) but cooler and more stable in the east (Figure 12.2) over the cold ocean currents (e.g. Canaries, Benguela and Peru Currents) = Tropical Maritime (Tm) type.

Classification of major air mass types

The Swedish scientist T. Bergeron proposed the first comprehensive classification of major air mass types in 1928, based on the source regions discussed above. Later modification included air flow trajectories in order to make some essential allowances for modification in the 'receptor' regions (Figure 12.1). It is now generally accepted that there is a main fourfold classification, based on two major categories, Polar (cold sources) and Tropical (warm sources). These represent semi-permanent stagnating anticyclones, as discussed earlier in this chapter, with each major type subdivided into maritime or continental (with their respective

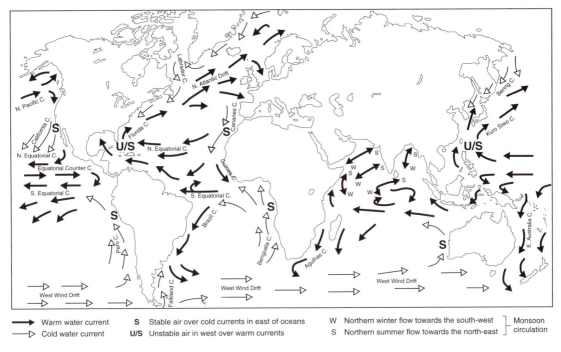

Figure 12.2 *The distribution of global ocean currents.*

homogeneous thermal and humidity properties). This four-fold classification is summarised in Table 12.1.

Air moving over warmer surfaces receives the suffix 'K' (kalt), implying heating from below with ELR steepening (see Figure 5.1), and it becomes increasingly unstable (suffix 'u'). Conversely, air moving over colder surfaces is suffixed 'W' (warm) implying cooling from below, with surface inversions (see Figures 5.1 and 5.3a) and increasing stability (suffix 's'). Hence, polar maritime (Pm) air moving equatorwards has a complete classification of PmKu, whereas tropical maritime air moving polewards is classified as TmWs. However, this basic and broad classification proves inadequate when applied on a regional scale. Consequently, various other sub-groups are in use in certain parts of the world, although they are mainly refinements of the four major sub-groups (Table 12.1). For example, the term equatorial air (E) is sometimes used to describe air of various origins and histories that has become stagnant over equatorial waters and has acquired excessive warmth and moisture (i.e. a 'super' Tm air mass). When such air is drawn into the summer (south-west) monsoon of South-east Asia, it is often termed equatorial monsoon air (Em), although the reverse winter circulation occasionally acquires Pc characteristics, with severe cold spells in northern India and many deaths from hypothermia. Similarly, Arctic (A) or Antarctic air masses are 'super' Pc types, which are excessively cold, dry and stable with very pronounced surface inversions.

Table 12.1 *Air mass classification*

Major group	Sub-group	Source region	Source properties	Airflow modifications
Polar (P)	Maritime (Pm)	Oceans north and south of latitude 50°	Cold, moist, unstable	Moving equatorwards, increasingly unstable with ELR steepening (Ku) and thunderstorms
Polar (P)	Continental (Pc)	Arctic regions, Antarctica, North America and Eurasia	Very cold, dry and stable (surface inversions)	Moving equatorwards over the sea, increasingly unstable with ELR steepening (Ku) and thunderstorms
Tropical (T)	Maritime (Tm)	Subtropical oceans	Very warm, moist and unstable in west (warm currents and ELR steepening); cooler and stable in east (cold currents and surface inversions)	Moving polewards, increasingly stable with advection fog (Ws)
Tropical (T)	Continental (Tc)	Low-latitude subtropical deserts	Very hot, dry and unstable in summer (ELR steepening) with dust storms; cooler, dry and more stable in winter	Moving over the sea, acquires moisture and prevailing instability favours thunderstorms

Air mass boundaries and frontogenesis

Neighbouring air masses are separated by gently sloping transition zones, which were termed fronts by Norwegian meteorologists in their pioneer research during the First World War (analogous to the military dispositions and activities of the war zone). The fronts were seen as comparatively narrow zones (perhaps 1000 km in width) of thermal discontinuity (i.e. baroclinic zones) in air mass properties. They became known as belts of sharp meteorological gradients, which were recognised as distinctive factors in the origin of middle-latitude disturbances (see Chapter 14).

The juxtaposition of air masses at a front has to be understood in a three-dimensional model (Figure 12.3), and there may be confusion due to the use of terms. The front is

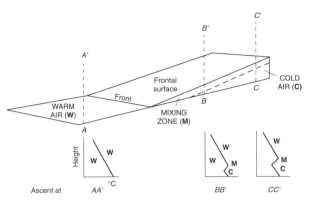

Figure 12.3 Schematic structure of a front

clearly a transitional layer about 1 km in thickness, sloping very gently (with a gradient usually between 1:25 and 1:300) up towards the cold air side. Within this layer, there is turbulent mixing between the two air masses. The term 'frontal surface' is sometimes (loosely) applied to this transitional layer, and where this 'surface' intersects the ground, it is termed the front. However, because of the very gentle frontal slope, this is in reality one edge of a belt of considerable width (up to 200 km, depending on the gradient) with the transition or mixing zone (Figure 12.3) being conventionally placed in the cold air. For convenience, these intersections are represented as lines on the synoptic (weather) map, with warm fronts indicating warm air replacing cold air on the ground, and *vice versa* for cold fronts. The schematic temperature/height graphs drawn from points A, B and C in Figure 12.3 illustrate the composite nature of the frontal structure. At points B and C, the mixing zone is represented by distinct frontal inversions when the adjacent air mass temperatures are very contrasted.

The process by which fronts form, or are intensified, is termed frontogenesis and is common in low-pressure regions where contrasting air masses converge (see Figures 10.9, 10.10 and 10.11). On the other hand, existing fronts weaken if the airflow pattern becomes divergent, with associated subsidence of air from higher levels. These processes accompany anticyclonic development, and fronts become inactive and eventually disappear (frontolysis) when pressure rises. It must be stated here that a frontal structure does not necessarily denote disturbed weather, since some fronts are relatively inactive. In this case, two adjacent air masses flow in parallel streams and no more than a weak cloud development betrays the frontal presence. A front becomes active when a convergent flow pattern results in the encroachment of one air mass upon the other, with the consequent ascent of the warm air leading to important latent heat and potential energy conversions (see Chapter 10 and Box 10.1). The frontal weather depends largely on the moisture and equilibrium properties of the rising air mass (see Chapter 5). For example, if this happens to be dry and stable (see Figure 5.2a), which is not common in middle latitudes, then the uplift has little effect and the front is weak, marked only by a belt of shallow (cumulus humilis) cloud. Conversely, when the warm air mass is conditionally unstable (see Figure 5.2c), a vigorous ascent is unleashed with cumulonimbus cloud embedded in the layer clouds and the risk of torrential rain and thunderstorms (see Figure 8.3). The detailed discussion of warm and cold frontal behaviour, especially associated with ana (ascending) and kata (descending) characteristics, will follow in the examination of mid-latitude weather disturbances in Chapter 14.

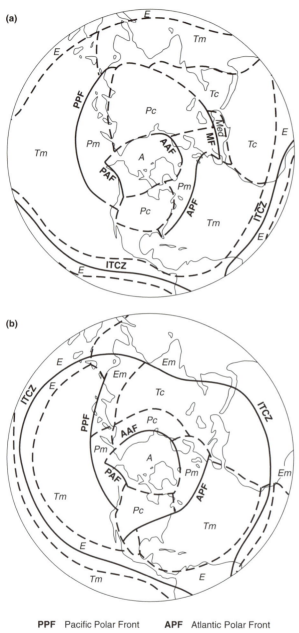

(a)

(b)

PPF	Pacific Polar Front	APF	Atlantic Polar Front
PAF	Pacific Arctic Front	AAF	Atlantic Arctic Front
MF	Mediterranean Front		

Figure 12.4 *Air mass regions and principal frontal zones in a Northern Hemisphere winter (a) and summer (b).*

This section concludes with an analysis of the distribution of air mass regions and principal frontal zones in winter and summer in the Northern Hemisphere (Figure 12.4). Two major frontal zones, the Polar and Arctic Fronts, are conspicuous over both these seasons, while a third one (the Mediterranean Front) occurs only during the winter season.

The Arctic Front

This is a distinctive baroclinic zone separating the ice and snow of the high Arctic regions from the more moderate polar/tundra environments to the south. The front roughly parallels the shores of the Arctic Ocean in both seasons, but the Pacific portion does move further south into north-west North America in winter as the high Arctic freeze is accentuated by the polar night (Figure 12.4a).

The Polar Front

This is a very well-known, active zone of frontogenesis in the Atlantic and Pacific Oceans (and indeed in the Southern Ocean mid-latitudes in the Southern Hemisphere). Figure 12.4 reveals that this front is quite variable, depending upon the seasonal distribution and extent of polar and tropical air masses. In winter, it shifts equatorwards (Figure 12.4a), when the Atlantic Front may extend into the Gulf of Mexico, to represent the juxtaposition of the cold North American (Pc) air mass and warmer Tm air mass. A secondary zone develops at this time in the central Pacific whenever the subtropical high there is split into two cells with converging air currents between them. In summer it contracts polewards, and two principal zones of Polar

Front activity occur over the middle latitudes of North America and western Asia/Japan in relation to the general weak meridional temperature gradient. Consequently, the frontal intensity is rather slight at this time of year, compared with the more vigorous winter activity, and Chapter 14 will relate these changes to the seasonal behaviour of mid-latitude depressions, which originate along the Polar Front.

The Mediterranean Front

This is only a winter feature (Figure 12.4a), when, at intervals, airstreams from Europe (Pm/Pc) and North Africa (Tc) converge over the Mediterranean Sea, bringing together air masses of markedly different temperature conditions. This convergence initiates and sustains frontogenesis, which can lead to cyclogenesis (see Chapter 14) and disturbed cyclonic activity with clouds, rain and gales. In summer, the Mediterranean basin lies under the influence of the STJ and the subtropical anticyclone (see Chapters 9 and 11), and the resultant dynamic divergence means that frontogenesis cannot occur, so hot, dry weather prevails.

Tropical discontinuities

Figure 12.4 also illustrates the seasonal variation in the location of the ITCZ in the Northern Hemisphere, which, as discussed in the Chapter 9 Case Study, was originally considered to be frontal in nature (the 1930s inter-tropical front or ITF terminology). Today, the ITF has been replaced by the term inter-tropical discontinuity (ITD) and, as illustrated in Figure 9.7, it exists (especially over West Africa in summer) only when very hot, dry Tc air converges with less hot, humid E air. However, as was mentioned in the case study, the ITD is no longer assumed to be thermodynamically similar to the Polar Front and no longer features in a discussion of frontogenesis.

CASE STUDY 12: The role of air masses in the secondary atmospheric circulation of Europe, North America and Australia

The variety of synoptic-scale weather and climate to be found within these three regions can largely be explained by the alternating residence of different air masses/airstreams, the fronts that separate them and the frontal disturbances (see Chapter 14) that develop over the regions or, more often, invade them from outside.

Europe experiences all the basic types of air mass classified in Table 12.1, but the sub-group 'Pm returning' must be added for a more complete description in Western Europe (Figure 12.5). Indeed, Western Europe is influenced by highly variable weather regimes, with a dominance of westerly (Atlantic) air masses, but it also receives

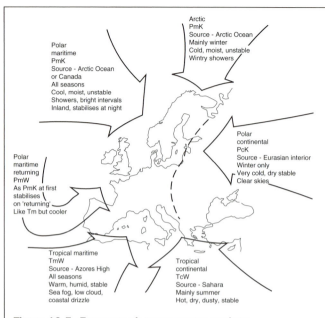

Figure 12.5 *European air mass source regions.*

more occasional airstreams with an eastern, continental origin. Conversely, Central and Eastern Europe experience more frequent continental air masses, with a reduced influence of maritime airstreams east of longitude 25° (roughly from Finland to Greece, as illustrated by the broken line in Figure 12.5). However, due to the west–east 'grain of relief' in southern Europe, the Mediterranean area is protected from continental air masses. In summer, particularly, it displays distinctive North African characteristics (see Case Study 11), with frequent outbursts of very hot, dry Tc (Saharan) air in the form of the Sirocco or Khamsin winds. However, in winter, the greater frontogenesis (Figure 12.4a) is responsible for active wave depressions in the Mediterranean area, and bursts of cold alpine/continental air to the coast (behind cold fronts), known as the Mistral or Bora winds (a unique airstream dominance, not experienced in other Mediterranean regimes, which are also motivated by the Genoa Low, see Chapter 9).

Figure 12.5 provides a schematic representation of European air masses in terms of their source regions and basic characteristics. Indeed, as mentioned above, the close proximity of tropical, polar, arctic, maritime and continental air masses leads to dramatic weather changes, especially in

Western Europe and the British Isles in particular. Figure 12.6a is a schematic representation of the potential British air mass variability related to anticyclonic and wave depression circulations (which operate singly). Of course, longer-lasting airstream control is associated with the development of blocking highs (see Chapter 11) and this leads to distinct spells of unseasonal weather. For example, the Arctic air mass develops under a weak thermal high over the Arctic Ocean. The resultant high-pressure ridge sends bitterly cold A air towards the British Isles, although it is modified in its passage across the North Atlantic (see Table 12.1) to reach the area as PmKu air (discussed earlier). The infamous British 'big freezes' are related more to semi-permanent winter highs (see Case Study 11) that persist over the snowbound Eurasian interior for weeks on end. The resultant Pc air mass spreads westwards, and during its passage across the North Sea, it becomes increasingly moist and unstable and can produce blizzard conditions on the east coast of Britain, especially in East Anglia and Kent. The best example of this 'big freeze' was the winter of 1962–63 in southern England, when nearly 1000 hours at or below 0 °C were recorded in January and February 1963.

Spells of hot, sunny weather (the so-called heatwaves) and droughts are related to poleward extensions of the STJ and the subtropical anticyclone during the summer season (see Case Study 11), when scorching dust-laden Saharan (Tc) air can reach the British Isles (Figure 12.6a) after some modification en route. The well-known

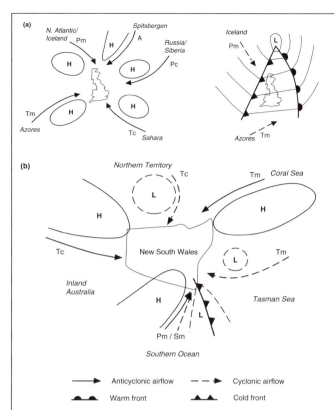

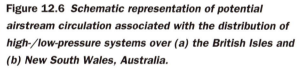

Figure 12.6 *Schematic representation of potential airstream circulation associated with the distribution of high-/low-pressure systems over (a) the British Isles and (b) New South Wales, Australia.*

British heatwaves/droughts of recent decades were associated with such persistent anticyclogenesis, especially 1975–76, 1984, 1989, 1990 and 1994. This increasing frequency might well be a symptom of global warming (see Case Study 3), but the massive circulation changes involved (see Part IV) seem to be quite independent of the very modest enhancement of the greenhouse effect experienced over recent decades. The invasion of Pm and Tm air masses (Figures 12.5 and 12.6a) is a much more regular occurrence, following extensions of Atlantic anticyclones, especially Tm from the Azores region. Then, the resultant TmWs airstream produces warm, humid, 'muggy' weather with advection fog shrouding windward coasts (especially in winter).

Conversely, in summer, greater airstream stagnation and land mass heating produce heatwaves terminating in violent thunderstorms (see Figure 8.3). TmWs airstreams frequently occur in the British Isles as very short-lived warm sectors in wave depressions (see Chapter 14). Also, behind the cold front in such a weather system, PmKu airstreams dominate with cool, showery and blustery weather, although they are more stabilised when returning northwards as Pm returning air (Figure 12.5).

North America experiences similar air mass types to Europe (Figure 12.7) although the 'grain' of the high Rocky Mountains produces marked differences in their properties and incidence. For example, Pacific influences are confined to a narrow west coast zone (compared with the greater westerly spread of Atlantic air masses), whereas polar and tropical outbreaks can reach the subtropical and subarctic regions, respectively. Indeed, Pc/A air from northern Canada is occasionally experienced on the coast of Mexico and over northern Florida, decimating the sensitive citrus fruits in these regions (see Case Study 7). However, advancing Pc air does lose stability during its passage over the Great Lakes and, combined with humidity changes, leads to heavy winter snow accumulation in the so-called Lake Peninsula snow belt area (between Windsor and Toronto, Ontario, for example). In summer, the snow-free but waterlogged Canadian Shield generates a very mild version of quasi-Pc air, which provides a welcome relief from oppressive Tm air as it advances southwards.

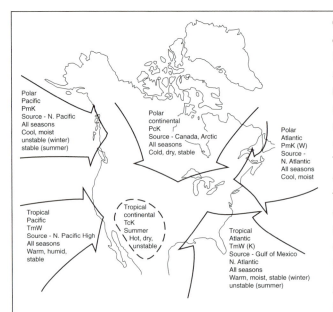

Polar
Pacific
PmK
Source - N. Pacific
All seasons
Cool, moist
unstable (winter)
stable (summer)

Polar
continental
PcK
Source - Canada, Arctic
All seasons
Cold, dry, stable

Polar
Atlantic
PmK (W)
Source -
N. Atlantic
All seasons
Cool, moist

Tropical
Pacific
TmW
Source - N. Pacific High
All seasons
Warm, humid,
stable

Tropical
continental
TcK
Summer
Hot, dry,
unstable

Tropical
Atlantic
TmW (K)
Source - Gulf of Mexico
N. Atlantic
All seasons
Warm, moist, stable (winter)
unstable (summer)

Figure 12.7 *North American air mass source regions.*

Tropical Atlantic air is typically TmWs in winter and produces advection fog (Figure 12.1b) and freezing rain (see Chapter 8) as it spreads northwards over the snow belt of Lake Peninsula. However, in summer, it dominates most of the United States east of the Rockies and pushes up into south-west Canada as a very oppressive, muggy airstream. Furthermore, it loses much of its initial stability (Figure 12.1a) as it moves across heated inland surfaces (due to lapse rate steepening, see Chapter 5) and is capable of unleashing violent thunderstorms, especially when lifted in frontal disturbances. Indeed, this uplift is very disastrous in spring and early summer, when marked contrasts occur along highly active cold fronts between Pc and Tm air, which lead to the formation of destructive tornadoes (see Chapter 16).

Pacific air is confined to the western seaboard by the Rocky Mountain 'blockage', and the Pmku type is similar to its European counterpart in origin, movement and characteristics (e.g. being particularly dominant behind cold fronts). Conversely, tropical Pacific air is inherently stable, and the prevailing upper-air (trade wind) inversion is strongly reinforced by surface inversions during the passage of the air mass (now TmWs) across the cold

Californian Current, which produces coastal (advection) fog (see Chapter 7, Figure 12.1b, Plate 7.2). The Rodgers and Hart hit song that hated California because it was cold and damp clearly refers to the fog-bound coastal belt and not to the inland areas, which develop Tc air (in summer particularly). However, the Tc air mass is confined to a small region of the south-western USA at this time of the year, although it is considered to be dry, stable Pacific air heated adiabatically by compression (see Chapter 5) as it descended the eastern slopes of the Rocky Mountains (see Figure 5.2a).

Australia (except for Tasmania) lies north of latitude 40° South and consequently does not experience Pc or even true Pm air masses at any time. Indeed, Pm airstreams (termed SPm or subpolar maritime on Figure 12.8 and Sm or southern maritime on Figure 12.6b) are greatly ameliorated during their passage equatorwards from the Southern Ocean. However, they still produce uncomfortably cool, moist conditions in the southern states (especially in Melbourne, Victoria), where their arrival is known as the infamous Southerly Buster. It is particularly uncomfortable in December when it accompanies a cold front/ subpolar trough moving quickly from west to east (Figure 12.6b). Ahead of the front, the 'norwester' drags in hot, dry and dusty Tc air over Melbourne (with temperatures in the upper 30s °C), only to be replaced in a matter of hours by the cool, damp and raw Sm air behind the front, with temperatures around 7 °C. This sudden air mass transfer represents the most dramatic weather change recorded on the Australian continent (e.g. it was responsible for the snowfall recorded for

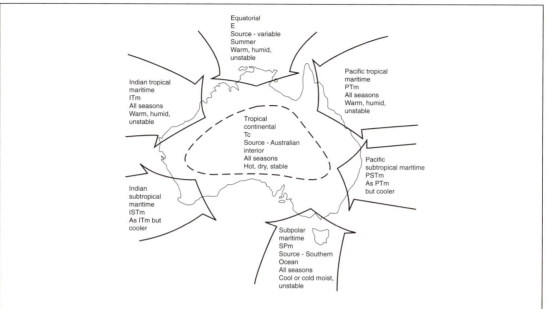

Figure 12.8 *Australian air mass source regions.*

the first time ever at Ayers Rock in July 1997).

There is not a great deal of difference in properties between Tm air, from both Indian and Pacific oceanic sources, and the E airstream that invades the extreme north in summer. Figure 12.8 illustrates that the changes between E, Tm and STm (subtropical maritime) air are minor, with a progressive cooling from equatorial to more subtropical oceanic environments. All these maritime air masses are moisture-laden and yield substantial rainfall when lifted by high relief (especially along the eastern seaboard, with a rapid decrease westwards, of about 15 cm km^{-1} in mean annual rainfall immediately in the lee of the escarpment) or in frontal situations. Tm air is particularly unstable in summer when it moves across a hot land mass, with severe lapse rate steepening and a potential for convectional rainfall.

Tc air dominates most of the interior of Australia, although it is considerably cooler and more stable in winter. The resultant thermal subsidence at this time reinforces the prevailing subtropical anticyclonic regime (associated with the 'return' of the STJ, as discussed in Case Study 11) and blocks the penetration of the peripheral maritime air masses. In summer, Tc air is very hot and very unstable, but it remains dry due to a very remote dew point (see Chapters 6 and 7). However, the ITCZ/humid air invasion into the interior at this time can produce occasional convective showers (and impressive waterfalls down Ayers Rock), which are most ineffective due to the excessive rates of vapour flux in the high-sun season.

Weather disturbances

The last chapter revealed that the middle-latitude juxtaposition of contrasting air masses (especially Pm/Pc and Tm air) has the potential for active frontogenesis and cyclogenesis. This activity produces potent depressions, which alternate with travelling anticyclones to produce the secondary circulation of weather disturbances in middle and high latitudes. At the same time, disturbed weather develops in low latitudes, especially in the summer/high-sun season, due to active ITCZ invasions and cyclogenesis, when resultant troughs/cyclones can also alternate with anticyclonic ridges to produce tropical secondary circulation or weather disturbances. This chapter covers:

- **the classification of weather disturbances**
- **the dynamics and characteristics of low-pressure systems (cyclogenesis)**
- **the dynamics and characteristics of high-pressure systems (anticyclogenesis)**
- **case study: the control of Rossby waves/PFJ on surface cyclogenesis/ anticyclogenesis**

Weather disturbances represent secondary or synoptic circulations, which are simply alternations of low- and high-pressure systems (see Chapter 9). They refer especially to those termed 'travelling' lows and highs, which are relatively short-term features embodied in (and hence disturbing) the global/general circulation discussed in Chapter 10. They develop from a wide range of atmospheric processes operating in low, middle and high latitudes, as explained in Box 13.1. The genesis of all these pressure systems was discussed in detail in Chapter 9, and this chapter will concentrate on the dynamics and characteristics of the weather disturbances involved, emphasising the complex control of upper-air systems.

The dynamics and characteristics of low- and high-pressure systems

All low-pressure systems have the same basic features and physical properties. For example, the rising air in the system (whether due to dynamic/Polar Front or thermal/ITCZ uplift) is cooling adiabatically, initially at the DALR and then, after condensation/dew point (when clouds and precipitation form), at the SALR (see Chapter 5). The resultant release of the latent heat of condensation is the primary energy cell for kinetic energy conversions and storm generation (see Chapter 10, Box 10.1). The rising air creates a vortex that spirals upwards due to the influence of the

---

Box 13.1

A classification of global weather disturbances (see Chapter 9)

1 The major low-pressure systems in middle/high latitudes are represented by wave depressions developing in all seasons along very active fronts (Arctic, Polar and Mediterranean frontogenesis, see Figure 12.4). Other (i.e. non-wave) depressions in these latitudes are independent of fronts and include orographic/lee lows and summer thermal lows.

2 The major low-pressure systems of low latitudes are represented by easterly waves and tropical cyclones/hurricanes, which develop from ITCZ cyclogenesis (see Case Study 9) only in the high-sun season.

3 The major high-pressure systems in middle/high latitudes are mainly those blocking highs developing from cold-air subsidence over Eurasia and North America in winter, i.e. the thermal or glacial anticyclones. Secondary sources are the travelling highs of low-index situations in the westerlies (Figure 11.6), especially on the retreating edge of an upper trough (see Figures 10.10 and 11.18).

4 The major high-pressure systems in low latitudes are those blocking dynamic highs developing from STJ convergence aloft and surface divergence in trade wind or subtropical anticyclones.

Coriolis force (see Chapter 10), which is subject to the control of latitude, namely it is a maximum influence at the poles. The rising air also leaves a vacuum, which leads to strong surface convergence, which is conducive to cyclogenesis. However, deep surface lows can intensify only when upper-air divergence is superimposed upon (and is greater than) the surface convergence (see Figures 9.2c, 11.17 and 11.18 point C). In middle/high latitudes, the PFJ represents the perfect 'scavenging' divergence aloft whereas the ETJ fulfils a similar role in the low-latitude monsoonal trough development (see Figures 11.15 and 11.16).

All high-pressure systems have the same basic features and physical properties, but these are the complete opposite to those in the lows discussed above. For example, air now sinks in the system (whether due to dynamic/subtropical or thermal/glacial/polar subsidence) and warms adiabatically by compression at the DALR over its entire descent. The resultant evaporation leads to cloudless, dry conditions. The sinking air spirals downwards due to the Coriolis force (see Chapter 10) and forms strong surface divergence/anticyclogenesis. However, anticyclones intensify only (into semi-permanent systems) when low-pressure convergence aloft (in the form of jet stream inflow, i.e. point A in Figure 11.18) exceeds surface outflow (see Figure 9.2d).

CASE STUDY 13: The role of Rossby waves and PFJ flow on surface highs and lows

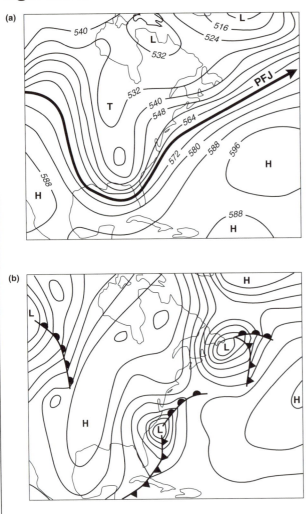

Figure 13.1 *The relationship between Rossby waves/PFJ and surface cyclogenesis/anticyclogenesis, eastern USA, 12/11/68.*

H High pressure   L Low pressure   T Trough of low pressure

The control of upper-air pressure/wind systems on surface cyclogenesis and anticyclogenesis cannot be emphasised enough, even though it has been discussed already (see Case Study 11, and Figures 11.17 and 11.18). For example, Figure 13.1 represents the surface and 500 mb pressure charts for the same date and time, and it is readily apparent that there is an obvious (if complex) association between the two charts. The two surface frontal depressions (labelled L and centred off Newfoundland and the Carolinas) are clearly located under the forward (western) limb of the upper trough, where PFJ scavenging/divergence aloft is evident (and vital for the intensification of cyclogenesis). Conversely, the surface anticyclone (H) is associated with the rear (eastern) limb of the trough, where PFJ convergence aloft is necessary for strong anticyclogenesis at the Earth's surface.

Weather disturbances of middle and high latitudes

Weather disturbances in these latitudes mainly represent the travelling highs and lows that dominate weather patterns in temperate and subpolar regions. Anticyclones and depressions develop due to favourable Rossby wave/PFJ situations (see Figure 11.18) and alternate with each other on a regular basis to produce the changeable weather regimes that typify these latitudes. More persistent (semi-permanent) blocking highs develop due to strong meridional flow/trough development in the Rossby waves (see Figures 11.6 and 11.8). However, the dynamics and characteristics of these systems were discussed in detail in Chapters 9 and 11 and will not be repeated here. Also, since anticyclones represent generally fine and settled weather, they are traditionally omitted from a discussion on weather disturbances. However, it must be remembered that they do have a 'nuisance' value in terms of the frost, fog and drought problems that characterise their persistence, and they should be regarded as relevant and potent disturbances in their own right. This chapter covers:

- **the general characteristics of wave depressions**
- **models of wave depressions**
- **cyclogenesis in middle–high latitudes**
- **life-cycle of a model wave depression**
- **non-wave depressions in middle–high latitudes**
- **case study: the characteristics and dynamics of warm and cold fronts**

The general characteristics of wave depressions (also known as a frontal wave and mid-latitude or extratropical cyclones) are revealed in Box 13.1 and typify the average conditions recorded during their passage. Of course, extreme situations develop from time to time, when their size, movement and life-cycle can be spectacularly different (e.g. the infamous British or 'Great' storm of October 1987, when its characteristics resembled a tropical cyclone in many respects).

Models of wave depressions

Weather maps have been produced on a regular basis in both the USA and the UK since the early 1870s, but they were basically pressure maps that clearly related cloudy, rainy weather with isobaric 'lows' and dry, settled weather with isobaric 'highs'. Indeed, the first ideas assumed that the lows had a thermal origin, although

---

Box 14.1

The general characteristics of a wave depression

1 The storm diameter is quite large and averages about 1600 km.

2 The storm travels quickly, averaging 1100 km day⁻¹ with a range of 300–1700 km day⁻¹.

3 The storm moves towards higher latitudes, steered by the meandering PFJ, which, through its function of rapid air removal (scavenging) is able to steer this movement.

4 The average pressure at the centre of the low is about 1000 mb, although a depression in the North Atlantic in December 1986 recorded 916 mb (a record low).

5 It is a short-lived system, since the greatest intensity of the storm is reached in two–three days, and the system normally fills/occludes within five days.

6 They dominate the winter season, when the Polar Front and PFJ are very active in middle latitudes. For example, the north-east coast of the USA averages twelve major wave depressions in January.

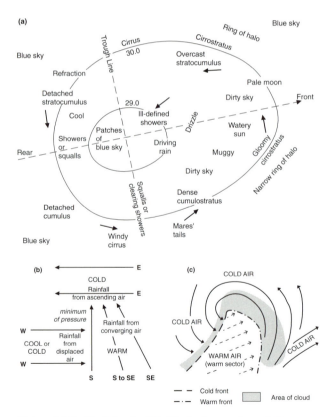

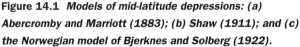

Figure 14.1 Models of mid-latitude depressions: (a) Abercromby and Marriott (1883); (b) Shaw (1911); and (c) the Norwegian model of Bjerknes and Solberg (1922).

Fitzroy and Dove had suggested the influence of converging air masses in the 1860s. Consequently, the first cyclonic model (produced by Abercromby and Marriott in 1883 and illustrated in Figure 14.1a) showed a simple, roughly concentric pattern of isobars with a descriptive labelling of the weather phenomena experienced. In 1911, air trajectories were included by Shaw into depression models (Figure 14.1b), which produced rather distinct discontinuities between a band of warm air from the south and cool/cold air from the west/north (the first suggestion of the cold front). However, it was left to the Norwegian meteorologists at Bergen during the First World War to complete the evolution of thought concerning the role of fronts in wave depressions (Figure 14.1c). At last, it was now possible to explain many of the surface weather features included in the earlier models (e.g. Figure 14.1a) and to integrate air mass, frontal and

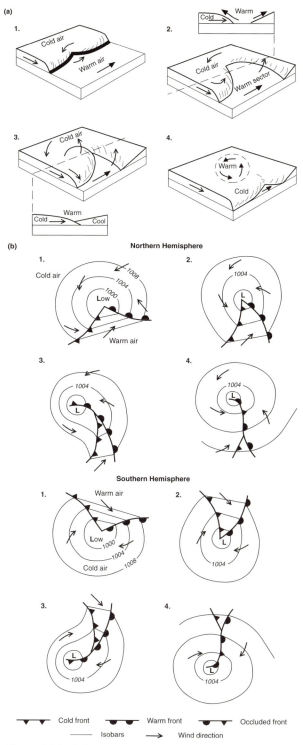

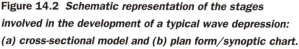

Figure 14.2 *Schematic representation of the stages involved in the development of a typical wave depression: (a) cross-sectional model and (b) plan form/synoptic chart.*

pressure development notions. Indeed, this Norwegian model has stood the test of time remarkably well, even though (like all models) it represents an idealisation and simplification. Many wave depressions resemble the model more or less at some stage in their development, although, as discussed later, the occlusion process (i.e. warm sector elimination) takes place much more quickly than the static models suggest (Figure 14.2).

Cyclogenesis in middle latitudes

The origin of cyclogenesis along the Polar Front in middle and high latitudes has already been explained in Chapters 9, 11 (case study), 12 and 13, when it became obvious that surface–upper air coupling was an essential feature of wave depression development. Of course, this association was unknown to the Norwegian meteorologists in the 1910s–1920s, who were unable to account for the rapid pressure fall at the tip of a warm sector in a mature wave. However, they clearly recognised the role of the Polar Front in the development of waves or undulations along the air mass boundary between cold air to the north (Northern Hemisphere) and warm air to the south. This control is still generally recognised as the dynamic origin of wave depressions.

However, it is not clear exactly how the wave is initiated, although the thermal/density contrasts of Pm/Tm air masses could provide the initial uplift (Figure 14.2a stage 1). This is then organised

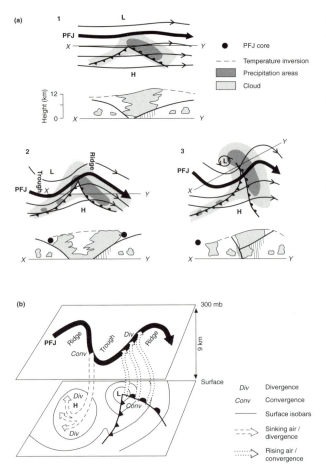

Figure 14.3 *(a) Stages in the development of an occluding wave depression; (b) Rossby waves/PFJ and surface pressure distribution.*

into strong cyclogenesis by upper-air divergence (i.e. Rossby wave/PFJ scavenging, Figures 11.18 and 14.3). Now the wave can develop into a mature, open form (warm sector), bordered by the warm front (on the advancing eastern edge) and a cold front on the retreating, western edge (Plate 14.1). This development is shown in a simplistic way in Figure 14.2 (stage 2 in both (a) and (b)), which represents an idealised cross-sectional model and a proposed synoptic chart. This figure also reveals that the 'life-span' of the wave/warm sector is relatively short due to the rapidly advancing cold front. This advance soon begins to undercut the warm sector and eliminate the warm air in the occlusion process (see next sections).

Life-cycle of a model wave depression

Figure 14.4 reveals the sequence of events in the development and decline of a wave depression, which is evident on consecutive synoptic charts of the North Atlantic for almost any week of the year (although the successive stages are more complex in reality). The initial stage (A) is a quasi-stationary portion of the Polar Front, with a warm westerly airflow to the south and a cold easterly flow to the north. However, both airflows may have a westerly component as long as the warm air moves faster than the cold air (with the necessary wind shear). The next stage (B) shows the beginnings of an eddy or wave superimposed upon the front. This is the 'birth' stage, with the first appearance of warm/cold fronts and converging rising air (releasing latent heat), now coupled with upper-air divergence (Figure 14.3). The central pressure begins to fall, and the incipient depression is now evident on a synoptic chart portrayed by two or three closed isobars (Figure 14.2a, stage 1).

By stage C, the system has deepened considerably and the amplitude of the wave is now enhanced by a well-developed warm sector and associated fronts (Plate 14.1). The depression has now reached its mature stage and has also progressed eastwards or

Plate 14.1 *Satellite view of a wave depression in the North Atlantic, NOAA5, 23/9/76.*

north-eastwards. For example, in the North Atlantic, it is common for such a low to be 'born' off Newfoundland and to have reached mid-Atlantic at the mature stage in a couple of days. Its further progress eastwards towards Europe brings about a fairly rapid achievement of stages D, E and F in two or three days. These three final stages are dominated by the elimination of the warm sector, which is known as an occlusion. It occurs because the more dense air behind the cold front tends to move faster than the warm front ahead of it and is destined (under most circumstances) to overtake it. This process starts at the 'tip' of the wave/warm sector and works its way along the wave, progressively forcing the warm air off the surface (stage E), giving protracted rainfall, especially if slow-moving. Eventually, two limbs of cold air mass become united and, in a fully occluded depression, the warm air is found only aloft (Figure 14.4, stages E and F).

As was discussed in Chapter 10 (Box 10.1), the juxtaposition of warm, light air and cold, dense air represents a high concentration of potential energy. The occlusion process clearly ensures that dense air underlies less dense air, which results in the conversion of potential energy into kinetic energy. This clearly enhances the vigour of the westerlies, including the PFJ, which, in turn, enhances its scavenging role and deepens the depression even further. Consequently, the stage of near-complete occlusion (F) is also that of maximum low-pressure deepening (and the formation of secondaries, explained below) and strongest winds. From this point on, the depression begins to 'die' as surface friction returns kinetic energy to heat (Box 10.1). Furthermore, the 'dwindling' uplift of air terminates the release of latent heat, which is vital for storm generation. The 'dying' depression fills rapidly, reducing the pressure gradient and slackening the winds, to be virtually unrecognisable on a synoptic chart. Plate 14.1 reveals the strongly organised cloud and rain belt of the occlusion/ occluded front, which is tightly coiled around the centre of the depression. Eventually, as the upper air loses its moisture supply, the clouds become more diffuse and the protracted frontal rain is replaced by occasional showers from patchy cumulus clouds.

Weather maps studied over a long period reveal a substantial number of complexities ignored by the simplicity of Figure 14.4 and the preceding section. These are mainly associated with the development of secondary lows within

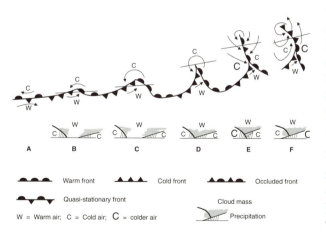

Warm front	Cold front	Occluded front
Quasi-stationary front		Cloud mass
W = Warm air; C = Cold air; C = colder air		Precipitation

Figure 14.4 *Idealised life-cycle of a wave depression.*

the circulation of the 'parent' or primary low. They may remain the minor partner in the system or may deepen to the same extent as the primary, giving a complex area of low pressure with two centres circling each other in a cyclonic direction. Sometimes this secondary low appears at the tip of what remains of the warm sector of a partially occluded primary low (the so-called triple point, where warm, cold and occluded fronts meet, i.e. in Figure 14.4, stage E). Occasionally, this secondary low moves away as a breakaway depression, while others deepen *in situ* to become the main low-pressure centre while the primary degenerates into a diffuse trough. Secondary lows often develop on the trailing cold front of an occluded depression and are termed cold front waves. When unstable, they can develop into a depression as vigorous as the primary (enhanced by PFJ divergence and scavenging aloft), and can develop a similar life-cycle (Figure 14.4). Repeated development of this kind gives rise to a so-called family of depressions (at stages of 'birth', maturity and 'death') strung out along the Atlantic Polar Front. This typifies the high zonal index situation (see Figure 11.7), with very unchangeable weather due to five or so members of the 'family' visiting regions in a week or so (with only a very temporary respite from each one due to an intervening but weak high-pressure ridge).

Non-wave depressions in middle latitudes

Chapter 9 discussed the origin of mid-latitude low-pressure systems that were not related to the Polar Front and, consequently, lacked the conspicuous eddy or wave/warm sector characteristics. Indeed, the origins were thermal/heat lows developing over inland areas in summer, and dynamic or orographic depressions forming in the lee of high mountain ranges. The detailed explanations need not be repeated here, and it will suffice to reiterate that thermal lows are the slow-moving thundery lows that develop over hot continental areas (coupled with upper-air divergence) and cause severe localised flooding. For example the disastrous floods over Poland and eastern Germany in July 1997, when the River Oder frequently burst through its dykes, causing horrific loss of life and unbelievable damage to crops and infrastructure. On a very localised scale, the convergence, spiralling air and latent heat release of thermal lows can lead to the spawning of horrendous tornadoes, although the most spectacular of these systems have origins along very active cold fronts (discussed in Case Study 12 and Chapter 16).

Lee depressions are large eddies that develop on the lee side of high mountain ranges, with associated dynamic convergence and cyclonic curvature and, initially, they appear 'anchored' to the barrier. As Chapter 9 explained, the origin is the vertical stretching of air descending the leeward slopes to form small, active depressions, which turn into unsettled, thundery systems. The Genoa Low is the best example of this disturbance, dominating the southern (leeward) slopes of the Italian Alps with northerly airflows. Regions in Alberta, Canada (in the lee of the Rockies) and

Patagonia in Argentina (in the lee of the Andes) are also favoured locations for lee depressions, resulting in violent thunderstorms (commonly at night in summer over Edmonton, Alberta) with destructive hailstones. Interestingly, the above three regions share a common hail problem, where pioneering research into hailstone suppression (through cloud seeding) is commonplace (see Case Study 8).

CASE STUDY 14: The characteristics and dynamics of warm and cold fronts

The above discussion on the development of wave depressions has highlighted warm and cold fronts as conspicuous, essential features of warm sector enhancement. Indeed, Figures 14.2 and 14.4 represent these fronts in an idealised way as simple zones of rising air due to thermal/density differences along the juxtaposition of contrasting airstreams. For example, at the warm front, 'light' Tm air rises over dense Pc air ahead of it, whereas at the cold front dense Pm air undercuts the 'light' Tm air ahead of it. Figure 14.5 illustrates these basic relationships along so-called 'textbook' warm (a) and cold (b) fronts. In the former

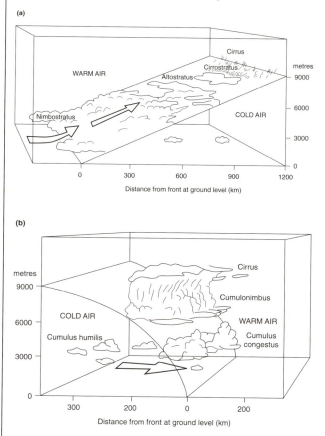

case, frictional drag exerted by the Earth's surface on the lower layers flattens the warm frontal slope but necessarily steepens it in the case of the cold front. Consequently, the ascent of air at the warm front (Figure 14.5a) and related cloud formation cover a much greater horizontal extent (up to 1200 km) than that of the cold front (200 km). Furthermore, the gradual gentle uplift of Tm air over the cold wedge (Pc air) forms cloud layers (stratus) increasing in height from lower nimbostratus, middle altostratus to upper cirrostratus (and cirrus at the leading outer edge), as illustrated in Figure 8.1 and explained in Chapter 8.

It is apparent that one-third of this cloud mass is likely to be thick enough to give the fine, prolonged rain or drizzle associated with layered clouds (see Chapter 8). Since the average warm front moves at about 15 m s^{-1} (although it can virtually stagnate when opposed by blocking highs), it means that the entire warm front cloud sequence will take about 18 hours to pass by a given location.

Figure 14.5 *Schematic representation of a 'textbook' (viz. ana) warm front and cold front.*

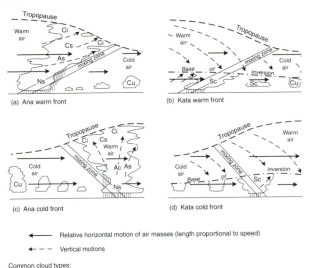

(a) Ana warm front

(b) Kata warm front

(c) Ana cold front

(d) Kata cold front

⟵ Relative horizontal motion of air masses (length proportional to speed)

⟵ − − Vertical motions

Common cloud types:
Ci = cirrus; Cs = cirrostratus; As = altostratus; Ac = altocumulus; Ns = nimbostratus;
Cu = cumulus; Sc = stratocumulus

Figure 14.6 *Ana versus kata fronts.*

Also, the warm frontal rain will persist for some 6 hours, giving evidence for the British folklore for the behaviour of warm fronts, namely 'rain before seven, fine after eleven'. Conversely, with cold fronts, the steepened frontal slope results in a much more vigorous uplift of warm Tm air, which leads to a pronounced vertical (cumiliform, as in Figure 8.1) cloud development, as illustrated in Figure 14.5b. Consequently, precipitation takes the form of torrential rain showers and hailstones, with thunderstorms and tornadoes characteristic of very active cold fronts (see Chapter 16). However, since the horizontal extent of the cumiliform cloud cover is about 1000 km less than the warm front cover, the duration of cold frontal cloud and precipitation is relatively short. Almost immediately behind the actual cold front zone, the cold dense Pm air begins to subside into flattened cumulus humilis clouds (Figure 14.5b), and this subsidence is usually accentuated by the approaching high-pressure cell or ridge.

It must be understood that these frontal generalisations mask a great variety in frontal behaviour. Indeed, in reality, fronts display ana (ascending) or kata (descending) features (compared with anabatic and katabatic winds, discussed in Chapter 16, which are illustrated in Figure 16.1b). The ana front is active at all levels, so there is a general uplift of warm air and extensive frontal cloud development, often up to the tropopause, in both the warm front (a) and cold front (c) situations. With a kata front, the necessary convergence occurs only in the lowest few kilometres, and warm air ascent (and flattened cloud formation) is confined to this layer. Above this narrow zone of uplift, a divergent or frontolytic condition (see Chapter 12) exists and the stable warm air subsides and dries out, with a subsidence inversion marking the base of the sinking air. This occurs in both warm and cold front situations, as revealed in Figure 14.6b and d. In Western Europe, ana warm fronts are more common than kata warm fronts, so the 'textbook' model (Figure 14.6a) is a reality. However, kata cold fronts are more common than ana cold fronts, so the 'textbook' model (Figure 14.5b) is quite a rare occurrence (with, fortunately, few tornadoes in evidence).

⟨15⟩ Tropical weather disturbances

Weather disturbances in these low latitudes are dominated by the semi-permanent anticyclogenesis controlled by STJ convergence aloft (see Chapters 9 and 11 and Box 13.1), which develops desert conditions in the subtropics. This is particularly so in the low-sun (winter) season, when the jet stream is located in subtropical regions and the ITCZ has moved into the hemisphere experiencing high-sun conditions (see Case Study 9 and Figure 9.7). At this time, persistent high-pressure cells produce generally settled weather compared with the greater activity and dominant cyclogenesis in the high-sun season, when the ITCZ returns over humid parts. The weather patterns are now dominated by very active low-pressure disturbances, ranging in scale from individual cumulus clouds to destructive hurricanes. As was the case in Chapter 14, since anticyclones in the tropics again represent generally fine and settled weather, they are traditionally omitted from a discussion of tropical weather disturbances. However, they do introduce crippling and devastating long-term droughts into the more humid parts of the region and, indeed, really disturb the atmosphere and the lives of the native peoples concerned. Reference to Chapters 9 and 10 will reveal the dynamics and characteristics of these subtropical highs, and this chapter will concentrate on the more important low-pressure systems observed in the humid tropics. It covers:

- the forms and features of subsynoptic weather disturbances
- synoptic-scale weather disturbances:
  1 easterly waves
  2 tropical cyclones
- case study: the influence of El Niño and the Southern Oscillation on tropical weather disturbances

Subsynoptic (or meso-scale) weather disturbances within the humid tropics are illustrated in Figure 15.1 and represent a series of convective systems ranging in size from 1 to 100 km. The smallest disturbances, with a lifetime of only a few hours, are simply individual cumulus clouds, which are deep convective cells (1–10 km in diameter) that develop from dynamically induced convergence in the trade wind boundary layer. Weak convective cells produce linear cloud patterns or so-called 'street' alignments of cumulus clouds, which occur roughly parallel to the airflow. With more pronounced convective mixing, the clouds can be organised into polygonal or honeycomb structures, especially when cold air moves over a warmer sea surface. The most vigorous convective towers are confined to regions where sea surface temperatures exceed 26 °C (i.e. where the dynamic convergence is accentuated by

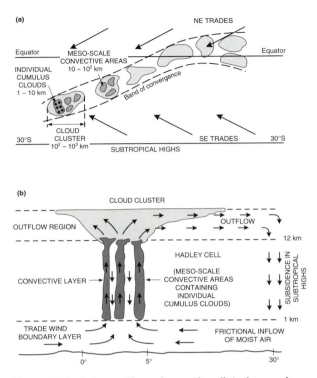

Figure 15.1 *Subsynoptic-scale weather disturbances in the humid tropics: (a) spatial distribution and (b) vertical structure (after Barry and Chorley, 1998).*

lapse rate steepening/free convection). They can extend over 20,000 m, with updraughts reaching 14m s⁻¹ and are accompanied by violent thunderstorms.

Groupings of individual cumulus cells produce meso-scale convective areas (Figure 15.1) up to 100 km across, where vigorous convergence and deep cumulonimbus clouds (thunderstorms) alternate with subsidence/divergence and cloud-free zones. Several of these mesoscale features then group into a cloud cluster (Figure 15.1), up to 1000 km in diameter. They represent concentrations of cyclonic vorticity at low levels and anticyclonic vorticity at high levels, with a reverse flow of strong subsidence and divergence between the convective areas. Cloud clusters are very common in the humid tropics, especially in the ITCZ season, when 10–15 such groups can occur per month and persist over several days. Even though these subsynoptic disturbances are small-scale and short-lived features, their latent heat release has a vital role to play in the destruction of the stabilising upper-air (trade wind) inversion, discussed in Chapter 5, Case Study (see Figure 5.3, plot B). This destruction is especially potent in Figure 15.1a, where seven cloud clusters dominate an area from the Equator to 25° latitude. Now, an extensive band of convergence exists that acts as a breeding ground for synoptic-scale cyclogenesis in the form of easterly waves and tropical cyclones.

Easterly waves

These synoptic-scale features develop in the broad convergence zone formed by an aggregation of several cloud clusters (Figure 15.1) within the ITCZ. However, the resultant vigorous convection takes place too close to the Equator, where the Coriolis force (see Chapter 10 and Figure 10.3) is too weak to generate circular atmospheric motions and major synoptic-scale disturbances comparable with the mid-latitude depressions discussed in the last chapter. Instead, the tropical convection becomes associated with pronounced wave-like undulations in the form of distinct low-pressure troughs and convergence zones athwart the trade winds (Figure 15.2) and

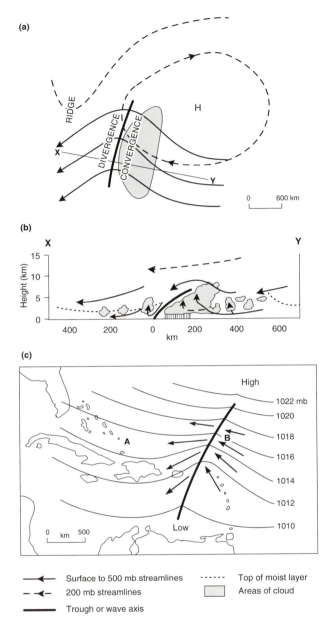

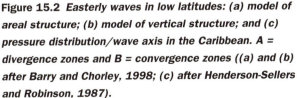

Figure 15.2 *Easterly waves in low latitudes: (a) model of areal structure; (b) model of vertical structure; and (c) pressure distribution/wave axis in the Caribbean. A = divergence zones and B = convergence zones ((a) and (b) after Barry and Chorley, 1998; (c) after Henderson-Sellers and Robinson, 1987).*

termed easterly waves. These troughs alternate with high-pressure ridges (divergence zones), with wavelengths between 2000 km and 4000 km. These waves have a life-span of up to two weeks and can travel westwards (in the trade wind flow) 6°–7° longitude per day, at a speed of 5–7 m s^{-1}, in the deep trade wind flow of the southern limb of the Azores and north Pacific subtropical highs.

Easterly waves are clearly identified from satellite images in the Caribbean and central Pacific Ocean, but their appearance is less regular than originally suggested. They represent shallow and weak troughs of low pressure extending from the Equator to the subtropics and are recognised on the weather map by a poleward tonguing or troughing of the isobars (Figure 15.2c). The origin of easterly waves is not clear, but they develop over tropical oceans like the Caribbean, where the trade wind inversion is weak or absent during the summer and autumn especially (due to the sub-synoptic destruction explained above). The absence of this stabilising upper-air inversion (see Figure 5.3, plot B) within the ITCZ allows the surface heating and lapse rate steepening to develop into unstable, freely convected systems, with considerable amounts of latent heat released due to excessive condensation (see Chapter 7). Extensive low-pressure troughs develop, which cannot be organised into cyclonic circulations because of their low-latitude location and associated weak Coriolis force.

A distinctive feature of easterly waves is the weather sequence observed before, during and after the passage of the trough axis (Figure 15.2a and 15.2b). Ahead of the axis, the trade wind inversion is particularly low and strong, prohibiting free convection and producing stable, clear weather (i.e. about 300 km ahead on Figure 15.2b and indicated by A in Figure 15.2c). This zone is also one of strong divergence as the air moves equatorwards and curves anticyclonically (in a high-pressure ridge), with vertical contraction of the air column and descending (at the DALR), drying and cloud-free air. Behind the trough axis, the inversion is much higher or indeed absent, allowing deep, free convection with cumulonimbus cloud in the unstable, humid air and heavy, thundery showers (i.e. at point B in Figure 15.2c). This point is a zone of strong convergence as the air moves polewards and curves cyclonically, with vertical expansion of the air column and ascending cooling/condensing air (at the SALR). Apart from this characteristic weather sequence, easterly waves appear to act as 'parent' vortices for the generation of tropical cyclone 'seedlings' (see next section), although only about 10 per cent of these disturbances survive to become mature hurricanes.

Tropical cyclones

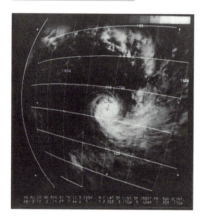

Plate 15.1 *Satellite view of a hurricane in the South Pacific, ESSA 3, 12/2/68.*

These weather disturbances are intense, circular low-pressure systems (Plate 15.1) with strong cyclonic circulations and vigorous airflows. They have a variety of different regional names, such as hurricanes in the North Atlantic and south-west Pacific, cyclones in the Indian Ocean/Bay of Bengal, typhoons in the north-west Pacific and even willy-willies in northern Australia (although this amusing name really should be restricted to active dust storms). The study of tropical cyclones has indeed been confused by these terminological discrepancies, which have limited the strict comparison of data on their intensity, frequency and movement. A useful classification, proposed by the World Meteorological Organisation in Geneva (Table 15.1), is generically based and includes tropical cyclones of all intensities.

Requirements for the development of tropical cyclones

Oceanic thermal energy represents the major energy source of a tropical cyclone, which develops a warm core in association with excessive rates of evaporation and subsequent condensation/latent heat release. High sea surface temperatures (in excess of 26°C and integrated to a depth of 60 m) are necessary for the required amount of lapse rate steepening and deep, free convection inherent in any tropical cyclone, which are initiated in close clusters of large cumulonimbus towers (discussed above). There appears to be a strong correlation between the seasonal location of the ITCZ

Table 15.1 *The World Meteorological Organisation classification of tropical cyclones*

Term	Wind speed
Tropical depression	up to 33 knots (17.1 m s^{-1}), Beaufort Scale, force 7
Tropical storm, moderate	between 34 and 47 knots (17.2–24.4 m s^{-1}), Beaufort Scale, force 8 or 9
Tropical storm, severe	between 48 and 63 knots (24.5–32.6 m s^{-1}), Beaufort Scale, force 10 or 11
Hurricane (or local synonym)	in excess of 64 knots (32.7 m s^{-1}), Beaufort Scale, force 12

Box 15.1

The general characteristics of tropical cyclones

1 They are generally small, normally in the range 200–600 km (or one-third the size of a typical wave depression, see Box 14.1).

2 The storms themselves travel quite slowly (*c.* 5–7 m s^{-1}) and move erratically, although they tend to recurve polewards around the western limb of a semi-permanent (blocking) subtropical anticyclone (i.e. to the south-east in Fijian hurricanes, as illustrated in Figure 15.3c).

3 The pressure at the centre of the low averages 950 mb, with a record low recorded in Hurricane Gilbert in the Caribbean (in September 1988) of 882 mb, which is considerably lower than that observed in wave depressions (Box 14.1).

4 The resultant pressure gradients are very steep (e.g. close to 1 mb km^{-1} in Hurricane Bebe in Fiji October 1972), which resulted in wind speeds approaching 100 m s^{-1}.

5 The storms require a long 'gestation' period in order for the latent heat release to reach maximum 'potency', and their greatest intensity is reached after 8–10 days of such energy release.

6 Torrential rain characterises these storms, averaging 25 mm h^{-1}, with a record 24-hour total of 1150 mm recorded in Manila, the Philippines.

7 The centre of the cyclone is characterised by a remarkable calm, clear 'eye', averaging 10 and 25 km across (Plate 15.1).

8 Cyclones occur at the end of the high-sun season (i.e. September/October in the Caribbean and February/March in Fiji, where three were recorded in a few weeks in 1983).

9 Their periodicity is quite irregular (Figure 15.3b), but their frequency has increased this century, which could be a symptom of global warming.

(and its thermal energy concentration) and cyclonicity. Evidence is provided by the absence of these disturbances in the South Atlantic and south-east Pacific, where the ITCZ is close to or north of the equator, respectively. Furthermore, there is clearly a distinct cyclone 'season', related to ITCZ control, which occurs in late summer/early autumn, when ocean temperatures are at their highest. For example, in the South Pacific the 'season' occurs between December and April, whereas in the Caribbean, it dominates the period July to October.

As was mentioned above, a pre-existing weather disturbance is required to act as a 'parent' vortex in order to release sufficient latent energy to break down the stabilising trade wind inversion. The necessary convective mixing can be provided by sequences of active cloud clusters (and 100–200 interconnected cumulonimbus towers) and the associate band of convergence (Figure 15.1), but it is usual for an easterly wave (Figure 15.2) to provide the 'parenthood' convection necessary for the generation of tropical cyclone 'seedlings'. Furthermore, as was discussed at the end of the easterly wave section, only about 10 per cent of these initial disturbances can develop into destructive, mature hurricanes. It is apparent that this high 'failure' rate in developing cyclones is due to other basic requirements not being evident at the time of seedling initiation. For example, a latitudinal control is associated with cyclogenesis rarely taking place near the Equator, where the Coriolis force is zero and the vertical component of planetary vorticity (see Chapter 10 and Figures 10.3 and 10.7) is absent.

This means that any semblance of organised balanced rotational motion (so vital to sustain a deep cyclonic circulation) does not normally occur within 5° of the Equator. Indeed, most of these motions only become organised enough polewards of 10° or 15° latitude. Also, tropical cyclogenesis requires weak vertical shear of the horizontal wind in the basic air circulation. With strong wind shear, the latent heat of condensation released by the developing convection (i.e. the primary energy cell or 'hot tower' in a tropical cyclone) will be

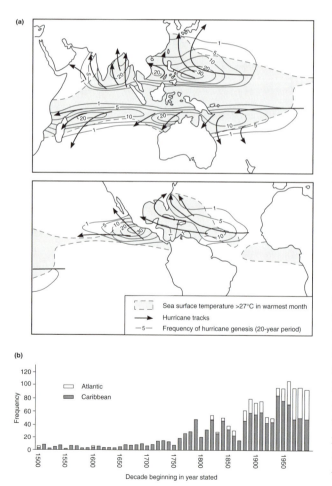

(a)

(b)

Sea surface temperature >27°C in warmest month
Hurricane tracks
—5— Frequency of hurricane genesis (20-year period)

120
100
80
60
40
20
0

Frequency

☐ Atlantic
▨ Caribbean

1500 1550 1600 1650 1700 1750 1800 1850 1900 1950

Decade beginning in year stated

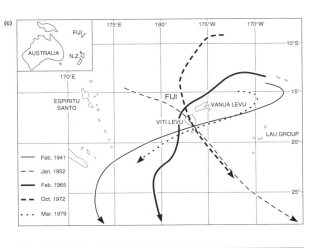

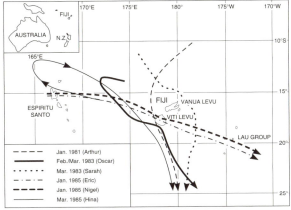

Figure 15.3 (left and above) *Tropical cyclones/hurricanes: (a) frequency of genesis and principal tracks; (b) frequency in the western North Atlantic, 1500–1990s; (c) frequency over Fiji Islands, 1940–1979 and 1980–1985.*

dissipated in the upper atmosphere. This is because the heat energy in these layers is now advected in a different direction from that in the lower layers, where this energy is released. Therefore, cyclone formation will take place only in latitudes equatorward of the STJ. Furthermore, strong wind shear is always present throughout the year in the western South Atlantic and central Pacific, so cyclones cannot develop in these areas. Indeed, the weakest tropical wind shear is found in the western North Pacific, where cyclogenesis peaks (Figure 15.3a), with about 30 cyclones developing over a 20-year period.

The final requirement for tropical cyclogenesis is associated with the mechanism necessary to couple the low-level convergence with a divergent flow above 12 km (the 200 mb pressure level) in order to sustain an intense free convection and storm generation. Such high-level outflow (at a greater rate than surface inflow, see Figure 9.2c) is normally provided by an upper-air anticyclonic cell or ridge. However, it also occurs on the eastern limb of an upper trough in the westerlies (point C in Figure 11.18), in the right latitude for the other basic requirements. The divergence aloft maintains the ascent and low-level inflow that is necessary to generate potential and kinetic energy continually from latent heat release (see Box 10.11).

These requirements are essential for the maintenance of cyclone intensity and as soon as one factor declines, the storm begins to decay. Degeneration occurs relatively quickly when the storm trajectory takes the vortex over a cool sea surface and especially over land, where friction increases and the supply of water vapour (the main energy source) is rapidly reduced. Rapid decay also occurs when cold polar air enters the system, or when the vital upper-level outflow becomes detached from the surface vortex, which now fills since the inflow at lower levels exceeds the divergence aloft.

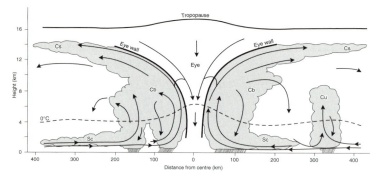

Figure 15.4 *Schematic representation of the vertical structure of a mature hurricane (arrows representing airflows involved).*

Meteorological features of tropical cyclones

The cloud structure in a tropical cyclone is cylindrical in form, extending from a low base almost to the tropopause. This cloud 'cylinder' widens out in its upper reaches and consists mainly of cumulonimbus towers massed together. Also, it is characterised by spiralling bands of stratocumulus and heavy cumulus, which enter at low levels, and cirrus/cirrostratus, which spread out at high levels (Figure 15.4). The 'eye' of the cylinder is a central well or funnel of subsidence and adiabatic warming of high-level air drawn down into the heart of the vortex. It is mainly cloud-free air (apart from some broken, thin cloud below 1500 m), which extends from the base to the top of the system, with a diameter of tens of kilometres.

Most of the energy of a large tropical cyclone is concentrated in a ring within 100 km of the centre, the winds attain maximum force in this zone. The strongest winds are found in a ring encircling the storm centre at a distance of about 24 km from the 'eye'. Inside the ring, wind speeds decrease rapidly to the relatively calm conditions of the 'eye'. Outside the high-velocity ring, wind speeds also decrease progressively towards the periphery of the storm. The central ring is characterised by the thickest cloud and heaviest precipitation, which can exceed 500 mm day^{-1}. For example, Hurricane Bebe in 1972 deposited 755 mm of rain between 23 and 24 October in Fiji as it passed over the mountains 27 km north-west of Suva, with pronounced forced convection (see Chapter 5).

The hurricane-force winds and excessive rainfall are coupled with storm surges (called *loka* in Fiji), which can reach heights of 15 m. *Loka* form over low-lying coasts when the abnormally low atmospheric pressure initiates a strong, positive sea surge (e.g. the sea level rises by approximately 1 cm for every 1 mb fall in pressure), drawing water up to some 30 cm and, with wind assistance, up to 4 m above normal sea level. Surges are particularly destructive along low-lying coasts, and their destructive potential is increased in estuaries when river floodwater meets and checks this sea water surge or when the wind impels the water against the coastline. Then, the water piles up to scour low-lying areas as a formidable wall.

Hurricanes usually move slowly, at 5–7 m s^{-1}, and initially the systems are steered from east to west in the trade wind flow (Figure 15.3c). Indeed, the tracks of most hurricanes are largely dictated by the deep circulation of the subtropical anticyclones in which they are embedded (Figure 15.3c). Hurricanes tend to recurve polewards around the western margins of these highs and eventually enter the circulation of the westerlies. They tend to decay as they move across the colder waters of high latitudes or over land, where the supply of latent heat is reduced considerably and the systems degenerate into weaker extra-tropical depressions. Alternatively, they can be regenerated in middle latitudes when they become incorporated into frontal depressions.

CASE STUDY 15: The influence of El Niño and the Southern Oscillation (ENSO) on tropical weather disturbances

When strong trade winds blow, frictional drag on the ocean transports warm surface water away from the source environment, and this divergence is driven by the Coriolis force. It then permits an upwelling of cold benthos water to the surface as in, for example, the Peruvian Current off South America (Figure 12.2 and 15.5a). Furthermore, the replacement of warm water by a cold supply has important atmospheric consequences. Lapse-rate steepening, instability and free convection over warm water are replaced by conductive chilling, a strong surface temperature inversion and stability.

In the tropical Pacific basin, the cold Peruvian Current is replaced at periodic intervals by the incursion of a weak, warm ocean current that flows south along the coasts of Ecuador and Peru. This disruption is traditionally referred to as El Niño, so named after the Christ child since it commonly occurs during the Christmas season. However, the El Niño event is now associated with the Pacific-wide climatic changes that are coincident with the more irregular occurrences of an exceptionally strong and warm current. These events are responsible for a more large-scale readjustment of the tropical atmospheric disturbances and oceanic regimes (Figure 15.5), which are now regarded as part of a long-term climatic variation in the Pacific basin known as the Southern Oscillation (SO).

The climatic readjustment between these alternating warm and cold occurrences was first described by Sir Gilbert Walker in 1928. Indeed, the so-called 'Walker circulation' represents the

normal state of the SO, when upwelling, cold deep water is conspicuous along the coasts of Ecuador and Peru (Figure 15.5a). This results in a 'normal' atmospheric–oceanic circulation mode with a distinct west to east longitudinal cell across the

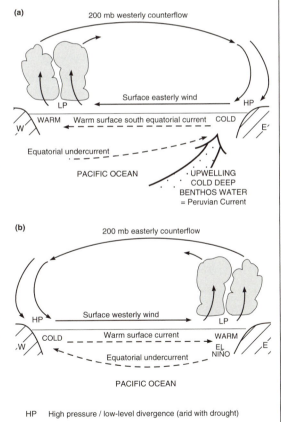

HP   High pressure / low-level divergence (arid with drought)

LP   Low pressure / low-level convergence (heavy rain with flooding)

Figure 15.5 *ENSO circulations: (a) normal Walker circulation and (b) El Niño circulation/reversal.*

Pacific. There now exists a typical trans-ocean pressure gradient at sea level of 5–10 mb between low pressure in the west and high pressure in the east (Figure 15.5a). Strong, surface easterly trade winds feed the low-level cyclonic convergence over eastern Australia, Fiji, India and Indonesia. This leads to the formation of large convectional disturbances, heavy monsoonal rainfall and flooding in these areas. At about the 200 mb level in the troposphere (Figure 15.5a), there is a distinct westerly counterflow feeding upper-air convergence over north-western South America. The associated subsidence and low-level anticyclonic divergence over Peru and Ecuador accentuates the aridity of these desert coasts.

It is apparent that this normal circulation mode is disturbed periodically during weak Hadley cell activity. Then, there is a replacement of cold water off South America by a southward flow of warm water, as equatorial currents, to about 6° S along the Peruvian coast. This replacement occurs at irregular intervals every two to seven years as El Niño. Then, coastal surface water temperatures suddenly rise by about 4° C, although El Niño normally takes from three to six months to reach local peak intensity after onset. Furthermore, the cessation of upwelling cold deep water and its nutrients causes economic chaos, with a massive fish kill and the loss of dependent sea birds (which is catastrophic for local guano industries).

El Niño is a very well-known phenomenon and, apart from its economic consequences, is responsible for torrential rainfall and massive flooding over the normally arid coastal deserts of northern Peru and Ecuador. This anomalous weather results from the weakening and reversal of the normal Walker circulation. Thermal low-pressure convergence develops over the warm El Niño waters off north-western South America (Figure 15.5b), which accentuates lapse rate steepening, free convection and precipitation totals. The normal

Walker circulation is linked to stronger Hadley cell activity, which increases the intensity of the surface trade winds and prevents warm water from flowing across the Pacific. This eventually terminates El Niño and, when the colder water is re-established in the eastern Pacific, the Hadley cell weakens and conditions are restored for the eventual, periodic return of warm equatorial currents as El Niño. The development and decay of El Niño characteristically take place over a period of twelve to eighteen months. It occurs roughly every three to four years, but the interval can vary between two years and a decade.

The reversed Walker circulation or ENSO episode (Figure 15.5b) is also responsible for pronounced anticyclogenesis over the south-west Pacific, including the eastern periphery of Australia. This development, and resultant westerly surface flow, is necessary to maintain cyclogenesis in the eastern Pacific, which in turn leads to an easterly counterflow at high altitudes, which is a product of the SO (Figure 15.5b). The persistent high-pressure subsidence in the south-west Pacific appears to introduce severe drought conditions into a normally humid tropical environment, including an impact on the Indian summer monsoon, which fails through a weakening of land–sea circulation systems. For example, the 1983 El Niño was a very intense episode and triggered massive drought conditions and forest fires in Australia, which were the worst recorded this century. Similarly, the 1987 drought in Fiji proved to be the worst one in at least the last century and has been attributed to the strong ENSO episode of that year. During this episode, a broad area of anomalously high pressure persisted over the islands continuously from April to October 1987. The resultant subsidence counteracted the normal orographic effect (discussed in the last sub-section and Chapter 5) and weakened the occasional frontal bands. Consequently, monthly rainfall totals recorded at Suva over the period were between 8 and 48 per cent of normal.

Conversely, 1988 experienced the disappearance of El Niño and the return of a strengthened, normal Walker circulation (termed La Nina, or girl child), which was responsible for unusually heavy rainfall in eastern Australia. At the same time, catastrophic flooding occurred in the Sudan, Bangladesh and southern China, with a spectacular millet harvest in the African Sahel for the first time for 20 years and above-average rainfall in India.

The current (1997) El Niño appears to be the strongest episode since 1983 and has been responsible for severe droughts in Indonesia (the worst for 50 years). Unfortunately, this aridity has been associated with massive forest-burning schemes throughout Sumatra and Kalimantan (largely by commercial palm oil and rubber plantation operators who are extending their lucrative properties), which produced dangerous smoke pollution levels throughout South-east Asia in September 1997. Kuala Lumpur and Kuching airports were closed, along with schools, and 50,000 people became ill in the choking smog. Two ships collided in poor visibility in the Straits of Malacca (with 29 deaths), and a Garuda Airlines airbus crashed in the smog near Medan (Sumatra),

killing 234 passengers and crew. The resultant ecological disaster is immense, with elephants, tigers and orang-utangs all under threat. Meanwhile, in the eastern Pacific, Hurricane Pauline caused unprecedented death and destruction in Acapulco (southern Mexico) on 9 October 1997. The unusual severity of this storm was partly attributed to El Niño peaking at that time, which caused above-average sea surface temperatures and associated instability. Southern California is expecting a return of the devastating winter storms and mud slides that were last experienced during the strong 1983 El Niño episode. This indeed happened during February 1998.

It is also apparent that strong teleconnections appear between these South Pacific ENSO events and extreme weather regimes around the globe, well outside the humid tropics. However, at the present time, we can only suggest that the occurrence of sea surface temperature anomalies seems to coincide with disrupted atmospheric patterns worldwide, and that Rossby wave positions (and the generation of surface lows and highs) appear to be very sensitive to such anomalies.

Tertiary circulations/local airflows

Part V concludes with a discussion of the smallest-scale atmospheric circulations, which represent the 'lesser fleas' and 'lesser whirls' of the Swift and Richardson quotes mentioned earlier. There is no general agreement concerning the scale limits of the tertiary or meso-scale (or simply local) circulations, although horizontal distances of up to 100 km are usually specified. However, since these circulations are often linked to terrain features such as mountain slopes and valleys, their actual dimensions obviously are very variable. Whatever their scale or origin, these circulations influence human activity and usually represent the difference between the regional weather forecast and the actual weather experienced in a given locality. This chapter covers:

- thermal circulations: land and sea breezes
- thermal circulations: slope and valley winds
- thermal circulations: urban airflows
- circulations in severe thunderstorms and tornadoes
- orographic circulations: lee eddies/rotors and föhn winds
- case study: classic sea breeze fronts in southern Britain

Land and sea breezes

These are common in summertime in coastal areas, when pressure gradients are weak (in anticyclonic conditions) and when the thermal contrasts between land and sea are at their greatest. There is a distinct, differential response of land and water to daytime heating and nocturnal cooling due to variations in specific heat capacity, heat transfers (vertical and horizontal) and vapour flux rates. Consequently, the land heats up and cools down more rapidly than the sea and, under slack gradients, a sufficient thermal contrast develops. This generates pressure differences (see Chapter 9) and a local air circulation (Figure 16.1a), which clearly reverses in direction as between day and night. The daytime sea breeze is directed from a relatively cool sea (high-pressure divergence) to a distinctly warm land (low-pressure convergence), and reaches velocities of up to 5–7 m s^{-1}. The oppositely directed land breeze at night, when less energy is available, rarely achieves half such speeds and is often reinforced by slope effects.

Figure 16.1a also reveals the presence of return flows at height (usually about 500 m in middle latitudes, but often twice that in the tropics) to complete the thermal

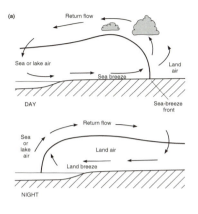

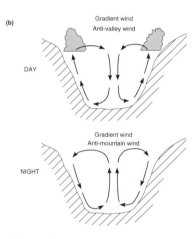

Figure 16.1 *Tertiary airflows: (a) sea–land breeze and (b) valley winds (after Oke, 1987).*

circulation. It also indicates a sea breeze front (marked by a line of cumulus cloud), which represents the juxtaposition of cool sea/lake air and the hotter land air. The progressive advance of this distinct front will be discussed (in a southern British example) in the case study at the end of this chapter. The cooling effect of the advancing sea breeze brings a welcome relief on some hot coastal lands, as is the case with the so-called Fremantle 'doctor' over south-west Western Australia. This important airflow tempers the afternoon heat and enriches the quality of the impressive wines from local vineyards (i.e. in the Swan Valley near Perth).

Slope and valley winds

Similar diurnally reversing airflows are also experienced in mountainous topography, where aspect and elevation combine to produce thermal circulations that resemble land and sea breezes in certain respects. During the day, net radiation surpluses (see Chapter 4) on sun-facing (adret) slopes create marked temperature/pressure gradients. These initiate airflows up the valley sides as an anabatic wind (Figure 16.1b, day situation and Figure 16.2), much favoured by hang-gliders on mountain summits and ridges. Indeed, the pressure gradients established will move air at ridge level from valley side to above the valley floor, which will sink to ground level and return the hang-gliders to terra firma (i.e. the anti-valley and valley winds of Figure 16.1b, day situation).

At night, the high valley sides cool first and temperature/pressure gradients are reversed. Cold air drains downslope under the influence of gravity as a katabatic airflow (see Case Study 5) and, to maintain continuity, air rises above the valley floor (as the so-called mountain and anti-mountain winds in Figure 16.1b, night situation, and Figure 16.2). This circulation (up to 5 m s^{-1}) is typical of clear nights and related long-wave radiation loss (see Chapter 3) with little wind, even in areas of quite modest local relief (*c.* 100 m). Katabatics are responsible for valley inversions and frost/fog pockets (see Case Study 5 and Chapter 7), which are generally avoided in the siting of settlements and sensitive crops/orchards. For example, in the so-called 'Garden of England' in Kent, inversion/frost hollows are avoided by hop growers, who prefer the thermal 'belt' between the more exposed mid-slopes and the inversion 'lid' (see Chapter 5, Figure 5.1).

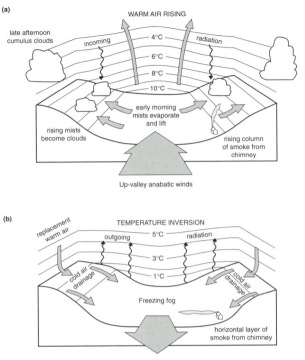

Figure 16.2 *Mountain and valley winds (after Oke, 1987).*

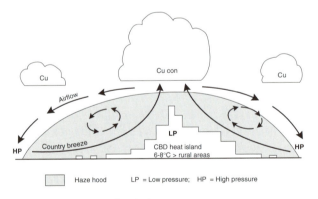

Figure 16.3 *Urban circulations.*

Urban airflows

Urban areas tend to be warmer than their rural surroundings for two basic reasons. First, the urban fabric/morphology has the capacity to absorb and store appreciable amounts of solar (net) radiation, particularly during calm, clear conditions in summer. This leads to a nocturnal build-up of heat over the city centre (the so-called Central Business District or CBD in Figure 16.3) and the creation of the summertime heat island (of large European cities), which can be some 6–8 °C warmer than the outskirts. Second, heat islands of similar magnitudes are produced by anthropogenic heat released from a variety of combustion processes, especially in North American and Russian cities/industrial regions in winter. Furthermore, the related smoke/water vapour 'blanket' reduces the nocturnal long-wave radiation loss over urban areas, which enhances the greenhouse effect (see Case Study 3).

Such urban–rural temperature gradients clearly develop distinctive thermal circulations on clear, calm nights, with cool rural air diverging from relatively high pressure and converging into the thermal lows over the CBD. This airflow is known as the 'country breeze' (Figure 16.3) and is common in such diverse cities as Frankfurt, London, Toronto and Asahikawa in Japan. Furthermore, this convergence leads to free convection over the CBD, which is enhanced by forced convection over the high-rise buildings. Consequently, cloud development and convectional rainstorms tend to be localised over cities, with rainfall assumed to be about 10 per cent more than in rural areas. However, it must be remembered that such dominant urban circulations are found only in cities located on flat terrain, which are free from the slope and valley interferences discussed earlier.

Circulation induced by severe thunderstorms and tornadoes

Chapter 8 and Figure 8.3 revealed the behaviour and characteristics of thunderstorms and the Chapter 12 case study discussed the spawning of tornadoes from such severe systems located along active ana cold fronts. In this section, the circulations only will be emphasised, to avoid repetition of the dynamics involved, and it is evident that the central up-currents and cloud/rain concentrations of such convective cells are coupled with compensatory subsidence in the surrounding, clear country air. Under very unstable conditions (aided, sometimes, by forced convection from ana fronts and relief features), the rising air may become organised into a convective 'chimney', about 1 km in diameter, fired by the latent heat of condensation. This system represents the incipient thundercloud (cumulonimbus), producing torrential rainfall, hailstones and electrical activity.

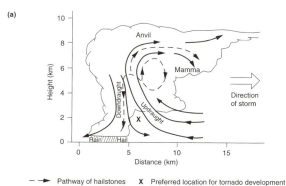

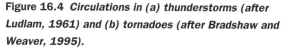

— → Pathway of hailstones   X Preferred location for tornado development

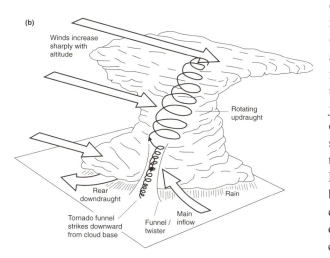

Figure 16.4 *Circulations in (a) thunderstorms (after Ludlam, 1961) and (b) tornadoes (after Bradshaw and Weaver, 1995).*

Figure 16.4a illustrates the circulation systems initiated in this developing convective cell, with these updraughts (reaching 10–20 m s^{-1}) holding ice crystals in suspension (formed through the Bergeron–Findeisen process, Figure 8.2). Eventually, these hydrometeors grow to exceed their terminal velocity (see Chapter 8, Table 8.1) and fall through the cloud as precipitation and drag down cold air by friction. This cold air downdraught (kept cold by raindrop evaporation/associated latent heat use) reaches the ground with the precipitation (Figure 16.4a). Here, it diverges (often against the main vortex airflow) to introduce cold, gusty outbursts with an upward 'jump' in air pressure. The juxtaposition of updraughts and downdraughts represents the mature stage in cell development, which is the time of most active circulation. Downdraughts are then assumed to become established throughout the cloud, which rains itself out in the final or dissipation stage. The typical life-cycle from initiation to dissipation is about one hour.

However, alternative views assume that the forward, vertically diverging surge of cold air at ground level assists in thrusting up the incoming air. This enhances a separate updraught, so that the storm is self-fed rather than self-destroyed, with new cells spawning on the advancing edge of the cell. The updraught is inclined or back-tilted (Figure 16.4a), which requires a strong wind shear in the upper air, normally associated with a cold frontal situation, which is particularly conducive to intense thunderstorm development. A frontal jet stream aloft (PFJ over the cold front) provides the necessary upper air scavenging/divergence (see Case Study 11) for the warm convective 'chimney' to 'draw' well and fire the system. Consequently, lines of thunderstorm cells (known as squall lines) are frequently found ahead of ana cold fronts (see Figure 14.6), triggered by downdraught 'scoops' from initial thundery outbreaks at the front itself. This development is particularly common in springtime/early summer in the central plains of the USA, associated with 'peak' thermal contrasts between Pc and Tc air along active, ana cold fronts (see Case Study 12), which lead to the formation of tornadoes (Box 16.1 and Figure 16.4b).

Box 16.1

Tornadoes

1 Tornadoes develop in severe thunderstorm circulations (called waterspouts over the sea).

2 They are very intense small-scale systems, commonly 100 m in diameter, but this can reach up to 1 km.

3 Wind speeds may approach 100 m s^{-1} as air converges rapidly into a central vortex, visible as a twisting funnel cloud, where the pressure fall ranges between 50 mb and 100 mb.

4 The funnel or twister originates at the base of the cumulonimbus (thunderstorm cell) and extends downwards towards the ground surface.

5 The rapid pressure drop and incredibly high wind speeds account for the destructive power of the tornado, which is capable of collapsing buildings and lifting heavy objects (like vehicles), albeit in a very narrow swathe.

6 The general conditions for tornado development are those discussed above, which lead to massive cumulonimbus (cnb) clouds, severe thunderstorms and squall lines. These include active ana cold front lifting of warm, moist unstable air, aided by jet stream scavenging aloft. This produces a rolling motion of air near the ground, which is turned into a vertical cylinder of rapidly rotating air within the cnb cloud.

7 Their actual origin appears to be related to an area just below the base of the thundercloud (point X in Figure 16.4a). Here, the interplay between the outspreading downdraughts and inblowing updraughts concentrates the cyclonic vorticity already present in the system.

8 In such conditions (Figure 16.4b), 150 of these destructive 'twisters' are reported (on average) in the Mississippi lowlands of the USA, mainly in spring and summer (when Tc/Pc thermal contrasts 'peak' at very active ana cold fronts).

Orographic circulations

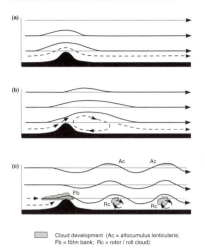

Cloud development (Ac = altocumulus lenticularis; Fb = föhn bank; Rc = rotor / roll cloud)

Figure 16.5 *Airflow over hilly terrain: (a) undisturbed laminar flow; (b) eddy streaming – one stationary lee eddy; (c) wave streaming – rotors and lenticularis clouds (after Corby, 1954).*

Chapter 5 revealed that forced (orographic) convection represents the vertical transfer of energy by eddy currents in the atmosphere, mainly associated with the obstruction to smooth (laminar) airflow (Figure 16.5a) imposed by mountainous regions. Indeed, these are completely unrelated to the thermally induced circulations discussed so far in this chapter. With all but a gentle airflow, an obstacle (ranging from an isolated building to a mountain mass) will cause a marked separation of flow, which departs from the pre-obstacle laminar flow. This causes a turbulent eddy on the lee side of the obstacle (Figure 16.5b) with a moderately strong airflow, although, with even stronger winds, the air descends on the lee side in a standing eddy. However, with still stronger winds, the lee eddy disappears and the air now develops lee wave oscillations downwind of the mountain crest (with a crest-to-crest wavelength comparable with the width of the mountain mass). Trains of up to six waves have been observed over 60 km distances and, sometimes, large turbulent eddies (known as rotors) form under individual wave crests (Figure 16.5c and Plate 16.1).

The final type of orographic circulation to be considered is the warm, dry wind descending the leeward slopes of mountain ranges, known as the Föhn or Foehn (Alps), the Chinook (Rockies), Santa Ana (southern California) or the Zonda of Argentina. Such an airflow can increase temperatures by 25 °C in an hour, reduce relative humidity by 50 per cent and produce wind gusts up to gale force. Their human impact is quite considerable, since they can cause avalanches, bush fires and rapid snow ablation (through sublimation, *viz.* snow pack to vapour), to encourage cattle grazing some weeks ahead of snow-covered pastures away from the föhn effect. The origin of this important wind was described in the context of the saturated adiabatic lapse rate (SALR) and stable equilibrium in Chapter 5 (Figure 5.2a and Plate 5.2). The classic explanation is related to forced convection and latent heat release following condensation on the windward slopes and the dry adiabatic descent (DALR) of stable air on the leeward slopes. However, very often the gain of

Plate 16.1 *Altocumulus lenticularis cloud near Palmerston North, New Zealand.*

temperature at the foot of these latter slopes is too great to be explained by these DALR/SALR differences alone, since pronounced föhn effects can occur with limited condensation/precipitation on windward slopes. Furthermore, a major query concerns the fact that warm air is descending on the leeward side, displacing colder, denser air. Consequently, the föhn wind should be regarded as basically part of a forced (orographic) circulation, that is it represents the lee eddy or lee wave discussed above, where windward ascent/heat release is not a prerequisite process.

CASE STUDY 16: Classic sea breeze fronts in southern Britain

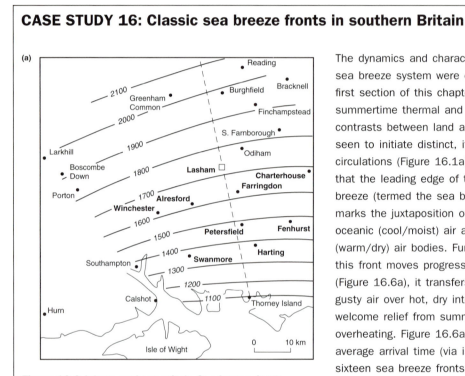

Figure 16.6 (above and opposite) *Sea breeze front: (a) arrival times averaged for sixteen fronts between 1962 and 1963; and (b) arrival of the front at Porton Down and associated weather changes, 24/7/59 (after Elliot, 1964).*

The dynamics and characteristics of the sea breeze system were examined in the first section of this chapter, when summertime thermal and pressure contrasts between land and sea were seen to initiate distinct, if local, air circulations (Figure 16.1a). It is apparent that the leading edge of the daytime sea breeze (termed the sea breeze front) marks the juxtaposition of contrasting oceanic (cool/moist) air and terrestrial (warm/dry) air bodies. Furthermore, as this front moves progressively inland (Figure 16.6a), it transfers cool, humid, gusty air over hot, dry interiors to bring a welcome relief from summertime overheating. Figure 16.6a illustrates the average arrival time (via isochrones) for sixteen sea breeze fronts in southern England (in 1962–63), where the advance over the 60 km from Thorney Island to Reading averaged 10 hours. Indeed, their arrival at Reading around 2030 hours provided invigorating fresh, breezy evening weather following an oppressively hot summer's day.

Figure 16.6b clearly illustrates the distinctive weather changes associated with the arrival of the sea breeze front at Porton Down, Hampshire (located on Figure 16.6a) at about 1800 hours on 24 July 1959. The initial dramatic weather change was an abrupt shift of wind direction from ENE (land air body) to SSW (oceanic air body), with a marked increase in velocity and gustiness. Air temperatures initially fell rapidly by 2.5 °C then decreased progressively to just below 20 °C by 2000 hours (representing a 7 °C decrease in 2½ hours). Similarly, the relative humidity jumped by 15 per cent as soon as the front arrives and then

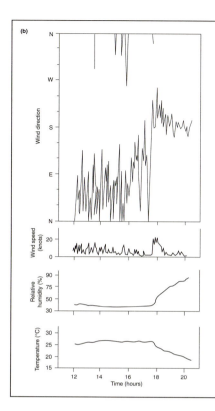

increased progressively by a further 35 per cent over the next 2½ hours.

A more modern study is listed in the Further Reading (which follows this case study), which plots the advance of a sea breeze front over the English Midlands on 30 June 1995. Indeed, this advance was a very rare occurrence, although conditions were ideal for its development and progress inland. For example, temperatures inland reached 31 °C by mid-afternoon, North Sea coastal water temperatures were 12 °C and a surface heat low/convergence developed in the region (with a central pressure of 1012 mb just south of Coventry). Also, a weak cold front was moving south (at about 12 knots) down the north-east coast of England during the afternoon (which provided an extra impetus for two sea breeze lobes). The sea breeze front moved inland at about 15 knots, and when it arrived in the Birmingham region at 1610 GMT, the wind gusted to 23 knots, its direction changed from 020° to 040°, temperatures decreased by 6 °C and dew point rose by 3 °C in a little under 30 minutes.

Key Topics for Part V

Secondary and tertiary circulations are superimposed on the global 'framework' of the primary (planetary) circulation and represent the prevailing weather regimes experienced in an area.

1 Air masses develop in polar and tropical locations (with continental and maritime characteristics), associated with the distribution of global high-pressure systems of thermal and dynamic origin. They move away from their source regions as dictated by the prevailing distribution of travelling highs and lows. These airstreams transport distinctive weather features to foreign receptor regions, with important modifications en route due to cooling/stability and warming/instability (e.g. TmWs and PmKu airstreams, respectively).

2 Air mass boundaries represent areas of frontogenesis, namely the Arctic/Antarctic, Polar and Mediterranean (winter only) Fronts, which vary spatially and temporally. Fronts are responsible for cyclogenesis and the formation of wave depressions in middle/high latitudes.

3 Wave depressions dominate the weather disturbances in middle latitudes and become intense when coupled with PFJ divergence/scavenging aloft. Their passage eastwards is occasionally blocked by thermal highs (in winter), or they can

alternate with the travelling highs, forming from the same low-index situations in the westerlies (but now associated with jet stream convergence).

4 Tropical weather disturbances range from subsynoptic systems (groupings of meso-scale convective cells) to the more important synoptic-scale easterly waves and tropical cyclones. Indeed, the former systems act as parent vortices for more intensive cyclogenesis, since they are responsible for distinctive free convection and latent heat release, which destroys the stabilising trade wind inversion.

5 However, only 10 per cent of easterly waves can develop into destructive cyclones, since other basic meteorological requirements are not evident at the time of seeding initiation. These include sufficient Coriolis force (normally 10–15° polewards of the Equator), weak wind shear and strong upper air-divergence/outflow.

6 Tertiary circulations are simply local airflows that are superimposed on the secondary systems. They are effective usually when pressure gradients are slack and local conditions become paramount. These include contrasts in land/sea heating/cooling rates (leading to sea/land breezes), anabatic/katabatic winds in areas of strong relief contrasts, and the country breeze into heated city centres. More intense localised heating/updraughts lead to severe thunderstorms, which can develop into tornadoes when the convective 'chimney' coincides with P.J.F. scavenging aloft (usually along active cold fronts).

Further Reading for Part V

Atmosphere, Weather and Climate. Roger Barry and Richard Chorley. 1998. Routledge.
A clear and concise account of secondary circulations, with an excellent non-technical discussion on mid-latitude and tropical disturbances.

Mountain Weather and Climate. Roger Barry. 1981. Routledge.
A unique account of orographic circulations, providing the detail not found in standard meteorological texts.

The Urban Climate. Helmut Landberg. 1981. Academic Press.
A clear, concise and non-technical account of urban circulations and associated climate characteristics.

Sea-breeze front reached Birmingham and beyond. J. F. P. Galvin. 1997. *Weather*, 52(2), 34–39.
An up-to-date study that plots the rare advance of a sea breeze front over the English Midlands, with sound meteorological explanations and clear reasoning.

 **Glossary**

The terms included are explained in the text but have been selected here because of their complexity, importance and/or confused general usage.

Adiabatic temperature changes refer to the changes of density and temperature that occur in a body of air as it rises and falls through the troposphere. Rising air expands and cools, since it uses up internal heat to supply the energy for the expansion process. Sinking air is compressed and warms up due to the increasing air pressure. It must be remembered that these changes within a body/parcel of air are isolated from the surrounding atmospheric environment (ELR), hence the term adiabatic (or self-contained/independent) rate of change. Rising air parcels cool at the dry adiabatic lapse rate (DALR, fixed at 10 °C km^{-1}) until dew point is reached. Thereafter, the saturated rate (SALR) takes over which is variable (averaging 5 °C km^{-1}), depending on the potential of the air for the release of latent heat from condensation, which counteracts the cooling by expansion. Subsiding air warms at the DALR unless evaporation is taking place. The relationship between the DALR/SALR and ELR controls the equilibrium condition of the troposphere (i.e. stable versus unstable air).

Air parcels are small bodies of air (normally represented by the dimensions of a heaped cumulus cloud) that rise through the atmospheric environment via free or forced convective systems (emphasised below). They cool adiabatically as they rise (DALR/SALR) and warm adiabatically (DALR) as they descend through the troposphere. Their thermal/density relationships with the surrounding atmospheric environment lead to the important equilibrium tendencies of stability and instability.

Airstreams are distinct flows of air from a donor source region, with polar/tropical and maritime/continental characteristics. They move in response to the prevailing pressure/circulation systems, especially blocking highs, which introduce a persistence into the airflow. Air mass source features are modified as the airstreams move into alien receptor regions and can be heated from below/made unstable (i.e. PmKu) or chilled from below/made stable (TmWs).

Albedo The coefficient of reflectivity of an object, expressed as a ratio between the total solar radiation received by a surface and the amount reflected (a decimal or percentage value). For example, the average albedo of the Earth is 0.34, or 34 per cent, but it varies according to the texture and colour of the surface. These variations are included in Table 3.2(a) and range between 80 per cent for virgin snow and 2 per cent for tarmac or bitumen.

Ana front is active at all times, since unstable conditions promote a pronounced uplift of warm air (analogous to anabatic winds) and extensive cumiliform cloud development at cold fronts, often to the tropopause, accompanied by torrential rain/hail showers. When coupled with jet stream scavenging, ana cold fronts can produce violent tornadoes, but these are less common in Western Europe (compared with the more frequent ana warm front).

Anthropogenic factors are associated with wide-ranging human activities that can modify natural atmospheric processes. These include deforestation, industrialisation and energy production, and are linked with important changes in atmospheric composition. Acid rain (from sulphate releases), the enhanced greenhouse effect (from CO_2 releases) and ozone depletion (from CFC degradation) all contribute to climate change, the latter two at the global scale.

Anticyclogenesis refers to the development of anticyclones, particularly the semi-permanent thermal/glacial highs and dynamic/subtropical highs that dominate the polar and subtropical deserts. Travelling (i.e. temporary) anticyclones alternate with wave depressions in middle latitudes, forming in low-index situations on the western edge of a Rossby trough, with strong PFJ convergence aloft.

Atmospheric black bodies are objects that radiate energy at the maximum possible rate per unit area and wavelength at any given temperature. A black body also absorbs all the radiant energy that is incident upon it. A true black body is rarely evident, although platinum soot approximates this ideal state.

Atmospheric coupling is a term used throughout the book to denote the essential three-dimensional linkages between surface and upper-air pressure systems. For example, surface lows can intensify into violent convergent cyclones or depressions only when they are coupled with strong anticyclonic divergence aloft. This is one of the reasons why easterly wave parent vortices in the humid tropics fail to produce destructive hurricanes.

Atmospheric equilibrium represents the degree of vertical motion that develops in rising air parcels due to the relationship between their internal (adiabatic) temperature changes (DALR/SALR) and those in the surrounding atmospheric environment (ELR). When parcels cool faster (i.e. the ELR is lower than the DALR and SALR), they become heavy/dense and resist further uplift due to free convection. This is stable equilibrium, and the reverse unstable condition occurs when parcels cool more slowly than the atmospheric environment (either absolutely or conditionally).

Atmospheric window is found between 8.5 μm and 14.0 μm wavelengths (terrestrial/long-wave/infrared radiation), where water vapour and CO_2 are ineffective absorbers/counter-radiators. Consequently, this energy is released into space and does not feature in the greenhouse effect. However, CFCs, released from a wide range of human products, are very effective absorbers at this particular wavelength. They are enhancing the greenhouse effect as well as contributing to stratospheric ozone depletion following their degradation into scavenging chlorine compounds.

Baroclinic zone is one of pronounced thermal/density contrasts within the atmosphere (compared with a barotropic zone, where no such contrasts exist). It is represented by a frontal zone, for example the Polar Front, which is associated with pronounced cyclogenesis (if the upper-air coupling is favourable).

Bergeron–Findeisen process is the growth of hydrometeors due to the coexistence of both supercooled droplets and ice crystals in a cumulonimbus cloud. Saturation vapour pressure (SVP) is low over the ice crystal, and this vortex attracts the flux of water vapour from around the supercooled droplet (high SVP). Consequently, the ice crystal grows by sublimation (to exceed its terminal velocity eventually and become precipitation), while the droplet disappears by evaporation (to maintain its SVP). This process represents the principle and practice of cloud seeding, using dry ice and silver iodide to stimulate SVP contrasts artificially.

Blocking highs represent persistent, stationary anticyclones, particularly in the westerlies, where they prevent the eastward movement of wave depressions, which are diverted away from middle latitudes. They can be extensions of either thermal/glacial highs or dynamic/subtropical highs, and are responsible for prolonged, settled, dry weather (with winter 'big freezes' and summer heatwaves/droughts). Surface blocking highs are coupled to upper-air trough convergence and are especially associated with jet stream behaviour.

Conservation of absolute vorticity involves a balance between the Coriolis force or planetary vorticity (PV) and the spin of the moving air body or relative vorticity (RV). Since Coriolis force is zero at the Equator and a maximum at the Poles, as an air mass moves polewards its relative vorticity must decrease to conserve absolute vorticity (and *vice versa* as it moves equatorwards). The responses are: reduced RV = negative spin, clockwise/anticyclonic curvature (Northern Hemisphere) towards the Equator, where RV now increases. So, increased RV = positive spin, anticlockwise/cyclonic curvature (Northern Hemisphere) towards the poles, where RV now decreases, so back to the reduced RV response. These responses account for massive wave oscillations in the westerlies of middle latitudes, known as Rossby waves with the meandering PFJ.

Conservation of angular momentum of a body (fixed mass) moving in a circle is proportional to its velocity and distance from the centre of the circle (i.e. the Earth's axis in the case of the atmosphere). So, as a body of air moves polewards, the associated decrease in radius must result in an increase in velocity, producing the vigorous westerlies of middle and high latitudes. Conversely, a body moving equatorwards experiences a radius increase and velocity decrease, producing the calm conditions (the Doldrums) in low latitudes.

Contour lines are used to represent upper air-pressure on a map (contour chart). Because of temperature/pressure complications with increasing height, it is more appropriate to depict the actual height of a selected pressure level (usually for 700 mb, 500 mb, 300 mb and 200 mb surfaces). Indeed, high heights correspond with high-pressure systems (i.e. expanding, warm air), whereas low heights reflect low-pressure systems (i.e. contracting, cold air). Consequently, upper-air contour charts are basically analogous to surface isobaric charts.

Convective systems represent the transfer of mass and energy within the troposphere by organised cells of unstable rising air. This uplift is associated with free convection (buoyancy due to surface heat accumulation/lapse rate steepening) or forced convection, which represents mechanical transfers due to strong frontal and orographic uplift of air parcels. These systems range in scale from individual cumulus clouds to more meso-scale groupings and finally a well-organised tropical cyclone circulation.

Cyclogenesis refers to the development of low-pressure systems, especially wave depressions along the active Polar Front and tropical cyclones along the ITCZ. These systems can intensify only when coupled to strong anticyclonic/jet stream divergence aloft. For example, travelling wave depressions in middle latitudes form in low-index situations on the eastern edge of a pronounced Rossby trough, with associated strong PFJ divergence (scavenging) aloft.

Dew point represents the critical temperature at which cooling air becomes saturated with water vapour (i.e. has reached its saturation vapour content, SVC). It is also known as the lifting condensation level (LCL) in rising air parcels, since any further cooling will result in the condensation of water droplets and cloud development.

Dynamic pressure systems are low- and high-pressure systems that develop through mechanical (i.e. non-thermal) processes. Dynamic lows form along active fronts (e.g. the Polar Front), where density contrasts facilitate the uplift of air and cyclogenesis. Dynamic highs are associated with jet stream convergence in the subtropics, where subsidence and surface divergence maintain the aridity of these regions.

Energy balance is also known as energy/heat transfers or non-radiative fluxes, where net radiation is partitioned into sensible (H) and latent (LE) heat at the Earth's surface. It operates at the regional or continental scale, where the partitioning reflects the resistance or otherwise to evaporation and evapotranspiration. For example, in pluvial South America, LE > H, whereas in arid Australia, H > LE. At the more local scale, sensible heat transfers can be divided into air and soil conductive processes (G), which again reflect the resistance to vapour flux together with soil thermal conductivity properties.

ENSO is the well-known abbreviation for the influence of El Niño and the Southern Oscillation. It refers to the periodic replacement of cold water/high pressure/drought off Ecuador and Peru by warm water/low pressure/floods. The former situation produces the normal or Walker circulation of water and winds flowing towards the cyclonic convergence zone over wet Australia, Fiji and Indonesia. The occurrence of warm water/low pressure off Ecuador and Peru is known as El Niño and completely reverses the above circulation, with cyclonic convergence and heavy rain over the usually arid coasts of these two countries. Conversely, high pressure and drought are experienced over the western Pacific, with droughts in Australia, Fiji and Indonesia, and even failure of the monsoon in India.

Evapotranspiration represents transpiration of water from the stomata, lenticles and cuticles of leaves and the evaporation of water in the soil zone adjacent to the plant's root system. Actual evapotranspiration is the 'real' loss of plant/soil–root water initiated by atmospheric factors and controlled by soil and plant characteristics. Potential evapotranspiration represents the water need of the plant where the supply of soil moisture is at all times sufficient to meet the demands of transpiring plants. It can be maintained only by irrigation, so that this rate of water flux is controlled mainly by the meteorological elements involved (especially net radiation and SVP), with stomata size/plant type the only other variables involved.

Frontogenesis is the term used to describe the process by which fronts form or are intensified and is common in low-pressure regions where contrasting air masses converge (i.e. the Polar Front).

Frontolysis is the opposite condition to frontogenesis and refers to the weakening/disappearance of fronts when airflow patterns become divergent, with associated subsidence of air from higher levels and anticyclogenesis.

Geostrophic wind is strictly defined as an Earth-turning wind, which is simply a balanced airflow (*viz.* parallel to the isobars) where the initial pressure gradient force (PGF) is counteracted by the Coriolis force of the Earth's rotation. However, this balance is not achieved in the so-called friction zone (within 600 m of the Earth's surface), where retardation destroys the wind velocity and the power of the Coriolis force. The airflow is now termed ageostrophic, with surface winds now blowing obliquely across the isobars as controlled by the more dominant PGF. Indeed, geostrophic winds are found only in the upper troposphere, remote from the Earth's friction, where the westerlies and jet stream concentrations flow parallel to the contour lines.

Greenhouse effect Confusion still exists with this term due to erroneous suggestions that the trapping of solar radiation is involved. Clearly, the greenhouse effect involves only outgoing terrestrial, long-wave, infrared radiation (LW↑) that is absorbed by trace gases (especially CO_2) and water vapour in the troposphere. When this absorbed energy is re-radiated back to the Earth's surface as returning

(LW$\downarrow$) infrared radiation, it can enhance surface temperatures and initiate global warming. The natural greenhouse effect is vital to prevent Planet Earth becoming as cold as the lunar surface. However, greenhouse gases produced from a wide range of human activities are assumed to be responsible for enhanced (if weak) global warming in the current century.

Hadley cell has been recognised for over 260 years to represent the thermally induced cell of air circulation between the Equator and the subtropics. On the equatorial limb is the thermal low-pressure zone (ITCZ) whereas dynamic anticyclogenesis (related to STJ convergence aloft) dominates the subtropical limb. These limbs are linked by surface trade winds and upper westerlies to maintain the cell in its traditional (simple) form. However, in reality, both the ITCZ and subtropical highs exhibit great variability in both their structure and location.

Heat island represents the concentration of heat within the urban area (CBD) due to its capacity to absorb and store solar (net) radiation, particularly during calm, clear weather conditions in summer. The resultant heat island, especially over European cities, can be 6–8 °C warmer than the surrounding rural areas. In winter, the solar effect is overtaken by the heat released by combustion within the urban area. which is enhanced by the smoke/vapour 'blanket' to reduce LW$\uparrow$. These winter heat islands (again of a 6–8 °C magnitude) typify North American and Russian cities.

Hydrometeors represent a generic term for all forms of condensation and sublimation of tropospheric water vapour, associated with cooling mechanisms aloft. Hydrometeor forms range from drizzle to hailstones and are controlled by the free/forced convective mechanism in operation and whether or not stability or instability exists.

Hygroscopic nuclei represent aerosols (like salt and sulphur trioxide) that have an affinity for water, which must exist in the troposphere to allow saturated air to condense into various hydrometeor forms. In their absence, air can become supersaturated and cooled to about –40 °C without sublimation taking place. Cloud seeding techniques involve the use of dry ice or freezing nuclei (AgI) to stimulate sublimation artificially through the Bergeron–Findeisen process.

Index cycle this is a schematic way of representing the development and decline of Rossby waves in the troposphere (usually over a four–six week period). The cycle represents the circulation changes from a zonal airflow (west to east) and high zonal index to meridional airflow (north–south and south–north) and low zonal index, as the wave oscillations become very pronounced over 3000–4000 km. At the same time, the PFJ meanders wildly and the respective convergence/divergence around a trough is coupled with surface anticyclogenesis (blocking highs) and cyclogenesis. Eventually, the waves are cut off, leaving isolated bodies of warm and cold air and a return to a high zonal index at the start of the next cycle.

Inversion of temperature represents a reversal of the normal ELR, when temperatures in the atmospheric environment increase with height. Inversions occur at the surface when boundary-layer air is chilled by conduction from terrestrial (LW↑) radiation on a clear, anticyclonic calm night (in winter especially). Upper-air inversions are found mainly in areas experiencing persistent subsidence warming from semi-permanent subtropical highs, with air descending 10,000–15,000 m. Inversions are responsible for very stable atmospheric conditions and serious pollution episodes.

Isobars are lines connecting places recording the same value of atmospheric pressure, after it has been reduced to sea level. An isobaric chart is also known as a synoptic chart or weather map, since it reveals the prevailing distribution of low-/high-pressure systems and frontal locations.

Jet streams represent high-velocity concentrations embodied in the upper tropospheric airflow. They are found in the westerlies, namely the PFJ at about latitude 45°, in the vicinity of the Polar Front, and the STJ at about 32° latitude in the subtropics. During the high-sun season, the ETJ is also recognised over Southeast Asia. Jet streams represent zones of highly concentrated kinetic energy transfers from enhanced potential/latent energy sources (e.g. the rising/condensing air along the Polar Front). Jet streams couple with surface disturbances, with PFJ/ETJ scavenging conducive to cyclogenesis (wave depressions/monsoonal circulations, respectively). Conversely, the strong convergence on the equatorial side of the STJ favours the subsidence and surface divergence of dynamic highs in the subtropics.

Kata front is generally inactive, since stable conditions are conducive to descending air (analogous to katabatic winds). Indeed, airstream convergence occurs only in the lowest few kilometres, and warm air ascent (and flattened cloud development) is confined to this layer. Above this narrow zone of uplift, divergent or frontolytic conditions exist, with a strong subsidence inversion accentuating stability. Kata warm fronts are rare in Western Europe, although Kata cold fronts are more common in this region but are not conducive to tornado development.

Kinetic energy represents the conversion of latent and potential energy into the energy for air movement. The PFJ is the best example of this energy conversion, due to the rising/condensing air at the Polar Front. Finally, kinetic energy is returned to heat following friction between the moving air and the Earth's surface/air molecules. Consequently, global winds would cease to blow within 13 days unless new conversions were constantly taking place.

Laminar flow represents a smooth flow of air over a featureless Earth's surface (e.g. an ice sheet or plain) compared with the turbulent eddies and lee wave oscillations that develop over mountainous terrain.

Lapse rate steepening is also referred to as a super adiabatic condition and represents an accentuation of the ELR by excessive surface heat accumulation on hot, sunny summer days. Under these conditions, the ELR can approach

14 °C km^{-1}, or more than twice its average rate. Furthermore, since the atmospheric environment is now cooling at a considerably faster rate than the DALR/SALR, extreme instability develops, leading to the formation of severe thunderstorms and even tornadoes (with the appropriate degree of upper-air scavenging).

Meridional airflow represents an advanced stage in the index cycle, when Rossby wave oscillations are most pronounced and a low zonal index exists. The winds now follow the lines of longitude (hence the term meridional) and are generally north to south or south to north. The PFJ now transports donor air into vastly different receptor regions, for example Alaskan (Pc) air can track as far south as Florida, devastating the sensitive tropical crops and citrus fruits.

Net radiation is defined as the balance between *all-wave* incoming and outgoing radiation, namely (SW$\downarrow$ – SW$\uparrow$ + LW$\downarrow$ – LW$\uparrow$). It represents the total energy available for all physical and biological responses in the Earth–atmosphere system. However, the biological processes of photosynthesis and respiration consume less than 1 per cent of the total net radiation (R_n) available. The bulk of the R_n is available for the surface–atmosphere energy balance and the partitioning into H, LE and G, discussed earlier in this glossary.

Occlusions represent the final stages of the life-cycle of a wave depression, when the warm sector becomes eliminated by the faster-moving Pm airstream behind the cold front. It starts at the 'tip' of the wave/warm sector and works its way along the wave, progressively forcing the warm air off the ground. Eventually, two limbs of cold air mass become united and, in a fully occluded depression, the warm air is found only aloft. With this diminishing supply of latent heat of condensation, the depression is deprived of its energy source and fills.

Relative humidity This is the most misused expression of atmospheric humidity, since it is erroneously attributed to a measure of low or high moisture levels. It simply represents the proximity of the air to saturation point and refers to the actual moisture level in the ambient air as a ratio (percentage) of the amount held when saturated at that temperature (namely 100 per cent). Clearly, it is directly proportional to temperature and decreases during the hot day and rises during the cooler night, when conductive chilling brings the air close to condensation/saturation point.

Thermal pressure systems are low- and high-pressure systems that develop through thermal processes. Thermal or glacial highs are shallow features that form over cold, snow-covered land surfaces due to excessive radiational–conductive chilling. Surface inversions are strong, which accentuates air stability. Thermal lows develop from surface heat concentrations and lapse rate steepening. The resultant instability leads to deeply developed cumiliform clouds, which intensify into a thundery low and even a tropical cyclone under the right atmospheric conditions.

Tropopause The troposphere is the lowest zone in the atmosphere, generally within

16 km of the Earth's surface. The top of this zone is known as the tropopause, which varies in height both spatially and temporally, being highest (18 km) at the Equator and lowest (8 km) at the poles, and higher in the summer months. Some 90 per cent of the water vapour and virtually all our weather systems occur in this zone, which is characterised by the variable ELR that controls the tropospheric equilibrium.

Upper-air scavenging A popular term used throughout the book to emphasise the critical role of jet stream divergence and its coupling to surface convergence/cyclogenesis, for example the ETJ and its influence on monsoonal intensification.

Wave depression The term is used throughout this book to denote the dynamic lows of the mid-latitudes, which form along active fronts (especially the Polar Front). This term is preferred to those used in American textbooks, which include frontal cyclone, mid-latitude cyclone and extra-tropical cyclone. The term depression is more realistic for mid-latitude lows, although some intense depressions can resemble tropical cyclones in their ferocity (e.g. the south of England storm in October 1987). The generic term 'cyclone' clearly has tropical implications, where the associated weather elements are always catastrophic.

Wind shear is used to describe a sudden variation of wind speed/direction with height and could well be a right-angled change in a horizontal airflow. It is clearly evident from differential cloud movements at varying elevations. Wind shear is a critical factor in cyclogenesis, since it dissipates the latent heat of condensation, which is not conducive to cyclone intensification. Consequently, tropical cyclones cannot develop in the western South Atlantic and central Pacific, where strong wind shear is present throughout the year.

Zonal airflow represents the initial stage in the index cycle before the Rossby wave oscillations develop, when a high zonal index exists. The winds now blow parallel to the lines of latitude, with a strong west–east component and weak transfers of mass and energy across the Earth–atmosphere system.

# References

Abercromby, R. and Marriot, W. (1883) Popular weather prognostics, *Quarterly Journal Royal Meteorological Society*, 9: 27–43.

Ahrens, C. D. (1991) *Meteorology Today*, West Publishing Co., St Paul, Minn.

Bach, W. (1972) *Atmospheric Pollution*, McGraw-Hill, New York.

Barry, R. G. (1969) The world hydrological cycle, in Chorley R. J. (ed.) *Water, Earth and Man*, Methuen, London.

Barry, R. G. and Chorley, R. J. (1998) *Atmosphere, Weather and Climate*, Routledge, London.

Bjerknes, J. and Solberg (1922) The life cycle of cyclones and the polar front theory of atmospheric circulation, *Geofysics Publ.* 3, no. 1.

Bradshaw, M. and Weaver, R. (1995) *Foundations of Physical Geography*, Wm C. Brown Publishers, Dubuque, Iowa.

Bridgman, H. (1990) *Global Air Pollution*, Belhaven Press, London.

Bridgman, H. (1997) Air pollution, in Thompson, R. D. and Perry, A. (eds) *Applied Climatology: Principles and Practice*, Routledge, London.

Chandler, T.C. (1969) *The Air Around Us*, The Natural History Press, New York.

Chang, J.-H. (1972) *Atmospheric Circulation Systems*, Oriental Publishing Co., Honolulu.

Corby, G. C. (1954) The airflow over mountains: a review of the state of current knowledge, *Quarterly Journal Royal Meteorological Society*, 80: 491–521.

Cotton, W. R. and Pielke, R. A. (1995) *Human Impacts on Weather and Climate*, Cambridge University Press, Cambridge.

Elliot, A. (1964) Sea breezes at Porton Down, *Weather*, 19: 147–150.

Galvin, J. F. P. (1997) Sea-breeze front reaches Birmingham and beyond, *Weather*, 52: 34–39.

Gloyne R. W. (1954) Some effects of shelter belts upon local and microclimate, *Forestry*, 27: 85–95.

Gribbin, J. (1988) *The Ozone Hole*, Corgi Books, London.

Hansen, J. *et al.* (1988) Global climatic changes as forecast by Goddard Institute for Space Studies three-dimensional model, *Journal Geophysical Research*, 93 (D8): 9341–9364.

Hanwell, J. (1980) *Atmospheric Processes*, George Allen & Unwin, London.

Henderson-Sellers, A. and Robinson, P. J. (1987) *Contemporary Climatology*, Longman, Harlow.

HMSO (1987) *Stratospheric Ozone*, Dept. of Environment/Meteorological Office Report, London.

Holmes, R. H. and Robertson, G. W. (1958) Conversion of latent evaporation to potential evapotranspiration, *Canadian Journal of Plant Sciences*, 38: 164–172.

King, K. (1961) Evaporation from land surfaces, *Proceedings Hydrological Symposium No. 2: Evaporation*, Ottawa, 55–80.

Lowry, W. P. (1967) *Weather and Life. An Introduction to Biometeorology*, Academic Press, New York.

Ludlam, F. H. (1961) The hailstorm, *Weather*, 16: 152–162.

Miller, A. and Thompson, J. C. (1970) *Elements of Meteorology*, Charles E. Merrill Publishing Co, Columbus, Ohio.

Morgan, M. D. and Moran, J. M. (1997) *Weather and People*, Prentice-Hall, Upper Saddle River, NJ.

Oke, T. R. (1987) *Boundary Layer Climates*, Routledge, London.

Oke, T. R. (1988) The urban energy balance, *Progress in Physical Geography*, 12: 471–508.

Oke, T. R., Yap, D. and Fuggle, R. F. (1972) Determination of urban sensible heat fluxes, *Proceedings International Geographical Union*, Montreal.

Palmén, E. and Newton, C. W. (1969) *Atmospheric Circulation Systems. Their Structure and Physical Interpretation*, Academic Press, London.

Pedgley, D. E. (1962) *A Course in Elementary Meteorology*, HMSO, London.

Petersen, J. T. and Flowers, E. C. (1977) Interaction between air pollution and solar radiation, *Solar Energy*, 19: 23–32.

Reading, A. J., Thompson, R. D. and Millington, A. C. (1995) *Humid Tropical Environments*, Blackwell, Oxford.

Rowntree, P. R. (1990) Estimates of future climatic change over Britain. Part 2, *Weather*, 45: 79–89.

Sawyer, J. S. (1957) Jet stream features of the Earth's atmosphere, *Weather*, 12: 333–334.

Sellers, W. D. (1965) *Physical Climatology*, Chicago, University of Chicago Press.

Shaw, W. N. (1911) *Forecasting Weather*, Constable, London.

Snow, J. T. (1984) The Tornado, *Scientific American*, W. H. Freeman & Co., San Francisco.

Terjung, W. H. *et al.* (1970) The energy balance climatology of a city–man system, *Annals Association American Geographers*, 60: 466–492.

Thompson, R. D. (1989) Short-term climatic change: evidence, causes, environmental consequences and strategies for action, *Progress in Physical Geography*, 13: 315–347.

Thompson, R. D. (1995) The impact of atmospheric aerosols on global climate: a review, *Progress in Physical Geography*, 19: 336–350.

Thompson, R. D., Mannion, A. M., Mitchell, C. W., Parry, M. and Townsend, J. R. G. (1992) *Processes in Physical Geography*, Longman, Harlow.

Thornthwaite, C. W. (1948) An approach towards a rational classification of climate, *Geographical Review*, 38: 55–94.

Ward, R. C. (1975) *Principles of Hydrology*, McGraw-Hill, London.

Whittow, J. B. (1984) *The Penguin Dictionary of Physical Geography*, Allen Lane/Penguin Books, London.

WMO (1956) *International Cloud Atlas*, Geneva.

Yap, D. and Oke, T. R. (1974) Sensible heat fluxes over an urban area – Vancouver, BC, *Journal Applied Meteorology*, 13: 880–890.

Index